# Daughters of Belgravia

## (Volume II)



Mrs. Alexander Fraser

Writat

This edition published in 2022

ISBN : 9789356396418

# Contents

# CHAPTER I
# LAST NIGHT.

"Trifles light as air
Are to the jealous confirmation strong
As proofs of Holy Writ."

"ALLOW me to congratulate you, Zai," Gabrielle says with a sneer.

Zai leans against the casement, idly toying with a spray of deep red roses she has just plucked from the trails that cover the wall hard by. She is very pale, and dark shadows underline her pretty eyes, and her thoughts are evidently far away, for she starts visibly as Gabrielle's voice falls on her ear.

"Congratulate me, and what for?" she answers rather bitterly.

Congratulations indeed! when her poor heart is so sore, her spirit so wounded by Carlton Conway's apparent defection last night.

"On your conquest of Lord Delaval," Gabrielle flashes out. "What a horrid little hypocrite you are, Zai! To think of how you spoke of him only yesterday morning and how you flung yourself at his head last night!"

"I don't understand," Zai murmurs, but her cheeks are quite flushed now and her grey eyes droop, for she remembers perfectly how, to pique Carl, she had flirted, as folks might think, with Lord Delaval.

"Zai! Zai! I thought you never told lies, and now you stand there in broad daylight uttering a monstrous falsehood."

Upon this, Zai bursts into an uncontrolable passion of tears, and flinging herself on the sofa presses down her face on the cushions.

Gabrielle attempts neither soothing nor scolding. To her such emotion is a display of childishness for which her hard nature has no sympathy. She rests calm and unmoved in her chair, languidly inhaling Eau de Cologne and occasionally sprinkling herself with a fragrant shower while she waits for the tears to subside.

"It seems very foolish spoiling your eyes by crying, Zai," she remarks at last contemptuously, when her not too great a stock of patience is, like the widow's cruse of oil, exhausted. "Of course I don't deny that Lord Delaval flirted with you as much as ever you could wish, and I suppose if you are engaged to him, it does not much matter if you *did afficher* yourself with him so shamefully."

"Gabrielle, you know I would sooner die than engage myself to that man!" Zai exclaims impetuously, dashing away her tears and sitting bolt upright.

"Child, you must surely be joking," answers Gabrielle, with a well-feigned accent of surprise, and with a quick uplifting in a curve of her dark brows.

Gabrielle is a rare actress by nature, and her vocation in life is the stage assuredly.

"Do you mean to tell me then that you are not engaged to him? If so you are certainly most indiscreet. All I know is, that if I descend to *afficher* myself before society with anyone, I shall take some man I like, and not one I was always professing to detest!"

"I *do* detest Lord Delaval!" cries Zai, in as shrill a tone as her bird-like voice can take. "I don't profess to detest him, but I detest him with all my heart and soul, and you know it."

"How on earth should I know it?" Gabrielle says sarcastically. "In fact I quite differ with you on this point; you may possibly fancy that you dislike him, but actions always speak so much louder than words that I am certainly sceptical."

"And pray what action of mine has shown any liking for him?" persists Zai, her eyes blazing angrily.

"Did your proceedings last night show any dislike? Instead of staying in the ball-room with the rest of the world, you prefer to remain outside. It was desperately dangerous and sentimental work that, Zai—only the Chinese lanterns and Lord Delaval's handsome eyes to keep you company, while you hung on his arm, and probably arrived at the conclusion that Lord Delaval is not worse looking than most of his sex!"

"Don't!"

There is quite a ring of pain in Zai's voice, and she gives a little shudder. The whole situation these last words bring so vividly before her is one she hates to realise, for she knows few would be charitable enough, and certainly not Carl, to give her credit for real dislike to such a rare-visaged Lothario as Lord Delaval, whose eyes, though their expression at times is hard and chilly as marble, can, when he desires, have an undeniable fascination in their sapphire depths, the bare outline of whose face is simply superb, and who looks what he is, an aristocrat all over.

The passionate looks Carlton Conway had given her have been her guiding star, and she believes that she would unhesitatingly follow their light into the deadly Styx itself, so it can be imagined how her very soul revolts as Gabrielle insinuates that she flung herself at Lord Delaval's head.

"Oh, Gabrielle, do you really doubt in your heart I would give anything I possess never to see Lord Delaval again and to be all right with Carl?"

"I don't care about going into possibilities," Gabrielle replies pettishly, "I prefer restraining myself to simple facts."

"Perhaps you will be less sceptical of my feelings if I explain a little about last night, Gabrielle," Zai murmurs deprecatingly. "You see I heard what Sir Everard said to you about Carl riding with Crystal Meredyth and looking 'awful spoons.' How those vulgar horrid words cut me through and through, Gabrielle! Then when we arrived, the first thing I saw was Carl waltzing with her, and—and—as if he really enjoyed it! I could not bear the sight of that, so when Lord Delaval proposed to go and see the illuminated grounds, I was thankful to go. After we had been out a little while I was anxious to come in, but he told me that Carl was engaged to Crystal—that he was obliged to marry some one who was rich, Gabrielle," and Zai flings herself down at her sister's feet and lifts up great pitiful eyes. "Instead of bullying me you ought to feel for me! I am heart-broken!"

"Heart-broken! You silly child, hearts are tough things and don't break so easily, I don't believe Carl Conway is going to marry that girl, but if he is, you must know he is a deceitful interested creature not worth thinking of. Well, what did you say to Lord Delaval in return for his information?"

"I only insisted on going round the garden by myself. I wanted to be alone with my wretchedness, and I wanted to call up courage to meet Carl face to face without betraying all I felt."

"Well?"

"Lord Delaval would not let me go alone, but I swear I forgot his existence even!"

Gabrielle gives a short unpleasant laugh at this.

"It is true, Heaven knows. We returned and were just going into the house when you and Sir Everard spoke about us—we were not a stone's throw from you, and of course every word you said fell out clear and distinct. I confess I was surprised at all I heard, as you know you did not speak the truth. However we won't discuss that point now. What I did hear made me resolve on an explanation with Lord Delaval at once. So I just told him frankly that I did not care for him and would never marry him!"

"In other words, you were amiable enough to reject him before he had the trouble of offering himself," Gabrielle says with a mocking smile.

"He had told me before that he loved me passionately, Gabrielle!" Zai murmurs with a hot deprecatory blush.

Her delicacy of character would not have let her reveal this except in defence of the seemingly fast conduct that has called down Gabrielle's sneers. And Gabrielle is well punished for her sneers—for this revelation of Zai's drives the colour from her cheek, and makes her writhe with jealousy.

"Very probably he did," she answers sharply. "Lord Delaval is a would-be monopoliser of women's hearts, and passionate love-making is one of the tricks of his trade. I don't believe there was a bit of genuine sentiment in all he said."

"I don't know, and I don't care if it was so. His protestations hadn't a feather's weight with me. And I never wish to see him again," Zai says quietly and truthfully.

"It never appears to strike you what people will say of last night. Society hasn't much romance in its composition. Society does not know, and would not credit that Zai Beranger wanders by day and night, blind to external influences, with a buckler girded on her heart on which is written 'Carlton Conway.' And if Belgravia cannot comprehend such high-flown sentiment, is it strange that I, born and bred amongst the *canaille*, with unlimited faith in the practical and matter-of-fact, and with a contempt for the foolish and the sickly romance of women, cannot help doubting and blaming you?"

"Blame has no effect on me," Zai says rather defiantly, with her little head erect. She is astonished and irritated at the cool condemnatory way in which it pleases Gabrielle to speak. It strikes her that there is too much presumption in it, and her really sweet nature, trodden on, like the traditional worm, seems inclined to "turn."

"But Lady Beranger is a slave to *on dits*, and she will lash herself into a fury if you don't carry out her scheme of marrying you to Lord Delaval, after your curious behaviour last night."

"It is mamma's fault, and not mine that it happened; she is always throwing Lord Delaval and me together, and the whole thing is hateful to me."

"Fiddlesticks! Mamma and lover being leagued, the odds are too much against you. You had better make up your mind to marry him; you will *have* to do so by-and-bye."

Lord Delaval's threat almost verbatim. Zai blanches with a sudden thrill of fear, and her heart gives a quick bound, but she says lightly:

"*Nous verrons!*"

"*Nous verrons!*" is the answer, and after a moment, Gabrielle goes on in studied accents: "I think it right to tell you, Zai, that I am resolved not to persuade you any more to marry Lord Delaval. I am a soldier of fortune, you know, and have to make my own way in the world; Lady Beranger deserves no tolerance from me, so I warn you, that I am going to try and serve myself, and if my interest clashes with anyone else's I won't yield an inch."

"In other words, Gabrielle, you give me notice that you are going in for Lord Delaval, yourself! I am sure I wish you *bon voyage* in your undertaking. I hope you will find the result, if gained, a happy one."

"I am not afraid, I never knew what fear was in my life. Cowardice in man or woman is the biggest crime in my eyes," Gabrielle says with a dare-devil glance.

"But," replies Zai, "why on earth should you consider it necessary to warn me of your project, I, who have no interest in the matter except to wish you happy?"

"Simply because I should wish the point made clear to you, so that you may not think me deceitful in the end. I owe the world—*your* world of Belgravia—nothing. But I have determined to take all I can gain from it by my woman's wit."

"Follow Trixy's example, and sell yourself to the highest bidder you can find in my world, then!"

"No one has ever bid high enough for me," Gabrielle cries bitterly, at the same time tossing her head with the proud air of a De Rohan. "Pariah, as I am, I have that which many of you Belgravians lack—the knowledge how to *live. Mon Dieu!* What a magnificent specimen of a *grande dame* I should make! Would that I were a peeress, and rich!"

Zai looks at her wonderingly, then she says quietly:

"I cannot think why people do not consider an inordinate desire for money sinful. It seems to me that money is at the bottom of every crime ever since our Lord Himself was betrayed for thirty pieces of silver."

"Why don't you preach all this to Trixy, then? She is practical in her greed for gold. You know all my rhapsodies may be purely theoretical."

"It would be a waste of time and breath to preach to Trixy. She has not a tenth part of your common sense, Gabrielle, and she cannot be held so accountable for her actions. Of course, mamma has literally coerced her into this awful match. She will endure the existence she has in prospect better than I should do, however. She won't think of Mr. Stubbs and his vulgarity while she has fine dresses and jewels. Sometimes I believe these things constitute her ideas of real happiness, do you know! But you, Gabrielle, are so different; if you pretend to lack a heart, at any rate you don't lack brains."

"No, I certainly don't," Gabrielle answers conceitedly.

Lack brains! Why it is on these very brains that she relies to bring honey and roses into her life, to get her luxury and ease, purple and fine linen, such as she loves actually quite as much as Trixy does: but has the *savoir faire*, or rather cunning, to keep her petty weaknesses locked up within the citadel of her own breast.

For a woman—and a young one—few could hanker more greedily after the flesh-pots and the silken attire of the children of Heth than this girl does.

To deck her ripe glowing beauty in the splendour of satins and velvets and soft bright hues, to see her long graceful throat encircled by the gleam of oriental pearls, her dusky braids crowned with a diadem of glittering brilliants, has been the dream of her life.

Ever since the old days when she loved to don a faded scarlet bow or a tarnished gilt brooch, to queen it over her sister *gamins*.

"By the way, Zai, I found out last night, that Baby has accepted old Archibald Hamilton! It was only by chance, as the little brat wants to keep the matter a secret from us for a while, I believe."

"Baby!" cries Zai, in amazement. "And yet I ought not to be surprised, for I might have read the news in Lord Delaval's face when he looked up from beside her at tea last night. I expect he likes *embarrass des richesses*, and is angry that even one of his worshippers should secede from her homage."

"It is no reason, because Baby gives her fat, dimpled hand to old Hamilton, that she should consider it necessary to close her heart to the fascination of her quasi lover!" says Gabrielle, with her Balzacian ideas, ideas that find no response in the pure mind of Zai.

"I can't stay chattering any longer, Gabrielle," she says hurriedly, and in the twinkling of an eye she is gone; and, as Gabrielle looks up surprised at her summary departure, she sees the tall figure of Lord Delaval slowly crossing the lawn towards the house, and guesses at once why Zai has disappeared in such haste. She bends forward, and, with wildly beating heart and tightly clenched hands, eagerly watches him.

Everyone who knew Gabrielle, sooner or later, asked themselves if she had a heart; and nobody amongst those most intimate with her, had yet been able to answer the question at all satisfactorily, excepting Lord Delaval.

But he did not seem to deem it worth his while to study her at all, though indirectly, and at all favourable opportunities, he let her be fully aware through the medium of his handsome eyes and his voice that he *knew* she had a heart, and that it was one he read like an open book and found remarkably interesting.

According to Dickens, there are chords in the human heart—strange varying strings which are only struck by accident, which will remain mute and senseless to appeals the most passionate and earnest, and respond at last to the slightest casual touch.

And so it is with Gabrielle.

She has reached over a quarter of a century.

Her nature is as passionate as that of a daughter of the south, and her early nurturing has been as wild and free as an Arab's; but no man's hand had struck the keynote of feeling until Lord Delaval put in an appearance on the scene.

He came, he saw, he conquered; and Gabrielle fell down at once, helplessly and hopelessly, to worship him.

# CHAPTER II
# FLIRTATION.

"What the years mean—how time dies, and is not slain,
How love grows, and laughs, and cries and wanes again,
These were things she came to know and take the measure,
When her play was played out so for one man's pleasure."

GABRIELLE'S cheeks grow crimson and her eyes glitter with pleasure, that for a little while they two will be alone, with no stranger to intermeddle with their joy, as she watches Lord Delaval approach nearer and nearer and finally step over the sill of the casement.

There is always a peculiar directness, an odd sort of intimacy in his manner towards her, whenever they are thrown alone together, that produces at once a most unconventional effect.

Now, as he walks up towards the sofa where she sits, the orthodox smile of greeting is lacking on his handsome face, the ordinary hand-clasp is unoffered, and Gabrielle does not even attempt to rise from her nest of downy cushions, while her face droops away a little from his gaze.

There is just a softer gleam in the big black eyes, a quick, nervous pressure of the even white teeth on the full, red underlip, and these are the only signs that she recognises his presence on the scene.

But Lord Delaval—confident and complacent—requires no spoken welcome. He has come in not knowing who he may find in the room, but finding Gabrielle, is ready, *faute de mieux*, to make love to her in the underhand way that does not compromise a man, and passes away an hour.

Ever since Baby's marriage to Archibald Hamilton had been hinted at by Lady Beranger, and he had suspected Zai's weakness for the popular actor, he had insinuated a passion, if he had not one, for Gabrielle. It may be that her evident liking for him, and her undeniable personal attraction, had touched him; but—probably it was only a selfish gratification he is given to seeking.

"I am so glad to find you alone. I wanted to see you so much," he says in a quiet outspoken fashion, that to a girl who hates what she terms the insincerity and shams of society is, in itself, fascinating.

"You wanted to see me, and you are glad to find me alone!" she repeats, then, to cover the nervousness his proximity always brings, she adds flippantly:

"Really, Lord Delaval, if Lady Beranger heard you she would drop at such a breach of the *convenances*."

"Possibly," he answers coolly, "but hang the *convenances*. Don't you know that there are times in every fellow's life when he comes into collision with the conventionalities, and either breaks them, or else risks being broken by keeping them? So long as I can run with my Juggurnauth, alias 'Society,' I am content, but I cannot throw myself before it and get mangled. Do you know I rather fancied I had a chance of finding you alone here, and so I determined to make chance a certainty?"

Gabrielle gives him a quick glance of surprise, while her heart throbs faster than it has ever done before in the six-and-twenty years she has lived.

Lord Delaval has often *looked* love at her—hinted at love, but he has never gone as far as this.

She has met him by appointment once or twice; still, nothing has been said to make her believe he really cared for her.

Now she reddens like a rose, and feels a nervous tremor run through her, and yet his manner is scarcely like a lover's. There is, in fact, nothing in what he says that could not pass as the ordinary talk of Society, yet the conversation seems lifted out from an ordinary atmosphere. They two, Lord Delaval and herself, are alone, and he talks to her just as if they were disembodied spirits. There are men occasionally in this world who have the power of bringing a woman they approach into direct contact with their own natures. They have a special gift of penetration, and one feels that in whatever relation one meets them, it is sustained by one's *real* self towards an equally *real* individuality on the other side.

Lord Delaval always makes Gabrielle feel this, and his intense manner adds to the feeling, but, with the supreme wilfulness of her nature, she refuses to yield to the magnetic influence he has over her without, at any rate, a struggle.

"You can have nothing to say to *me*, Lord Delaval, that all the world and the world's wife cannot hear. Are you mistaking me by chance for Zai?" she asks, carelessly, but she has no control over her features, and the excitement of his presence lends them a flashing, bewildering beauty, that positively dazzles him—*pro tem.!*

He fixes his deep blue eyes on her with an expression of fervid admiration, and her lids fall beneath the passion of his glance, but she lifts them bravely, and meets his gaze full.

"You really look as if you thought I did not mean what I say!"

"And no more you do, *ma belle*," he answers quietly. Outside the sun shines down furiously; the air is warm as an Indian summer. Up and down, up and down, the butterflies skim over the flowers, and a lazy rose-twig gives an inert tap on the window pane. Gabrielle does not reply. She feels shy, and as shyness is foreign to her, it is not only an uncomfortable, but a painful sensation.

"You snubbed Aylmer last evening," he says.

"Yes!" she answers laconically.

"But why? Did you forget how many good things he has to offer you? Most women would jump at such a match."

"*Soit!* but *I* don't," she answers indifferently.

"Of course not," he tells her. "I know you better than you know yourself—no one will ever know you as well as I do—and, still more, Gabrielle, no one will ever love you as *I* love you! No, don't start!"

For she rises from her seat, feelings of various kinds surge over her, and she clasps her fingers tightly together.

"Gabrielle, I have been longing to tell you this," he goes on, in a concentrated voice, which has a deal of suppressed passion in it; "I see no reason for denying myself the expression of what is strong within me. I don't want you to tell me that you love me, for I should hate to evoke from your sweet lips words that your *heart* doesn't force through them, in spite of *convenances*! I only want you to listen to me when, instead of dilating on the beauty of the weather, and so forth, I lay bare *my* heart to you."

Gabrielle believes he is laughing at her, and the belief lashes her into fury.

"Please, Lord Delaval, reserve your amusement for some one else. I am not of sufficiently elevated position for you to waste your breath on. Do you forget that Lady Beranger looks on me as a sort of social pariah, and almost a gutter-girl!" she flares out scornfully, her lips trembling, and looking doubly tempting in their wrath.

Perhaps Lord Delaval, with his worship for pretty things, feels their increased attraction, for as his eyes fall on them, his manner grows really more impassioned. He moves closer to her side on the sofa, but she averts her head, and piques him by a feigned coldness.

"I can't see your face, Gabrielle! And I *want* to see it while I talk to you," he pleads quite tenderly.

The tone touches her, not because she credits its sincerity, but because she has never dreamed that *he* could ever speak to her thus.

"Gabrielle, do you believe in affinities?"

"I believe in sympathy," she answers, wondering what he is going to say now.

"I am a firm believer in affinities, and don't believe in the possibility of love existing between two persons devoid of affinity. Tell me, Gabrielle! do you follow me at all?"

She makes a slight gesture of assent, but she doesn't in the slightest comprehend what he is driving at. No matter, he is close besides her. If she likes, she can touch him, and this is enough to put this impassioned child of Eve into a fever of delight.

"I don't believe that anyone can give another anything that does not belong to that other. He may withhold it to a certain degree, but it *must* be given in the end. Perfect love is when one meets someone to whom one can give all, and from whom one desires all."

"Imperfect affinities are all that most people in *our* world know of love, and, Gabrielle, Belgravia is horribly ignorant, do you know? Being so, they call a part of such and such a thing the whole, and demand allegiance of one's whole nature to a feeling that belongs to, and feeds but a small part of it! Now, Gabrielle—my beautiful, tempting Gabrielle! you and I have this in common, that we hate sham, and never pretend to fine sentimental feelings unless we possess them. Isn't it true?"

Lord Delaval bends over her till his face nearly touches hers, and he smiles conceitedly as he notices how rosy red the cheek near him grows by his proximity.

"I knew when I first saw you that you and I were exactly alike in our ideas and feelings. Somehow I felt it directly we spoke. I knew that you would never give to any man that which was not his—for you are dreadfully proud and cold and hard at the core, and when I found out, a day or two ago, that unconsciously I had learned to love you—do you hear me?—to love you with my whole being—when I found out that nothing short of an entire surrender of your soul—*of yourself*—would satisfy me, I trembled at the vision of bliss or torture that possibly lies before me—look at me, Gabrielle!"

There is a quiet command in his voice which she never attempts to resist. To everyone else sharp, caustic, cold, and full of sneers, to this man she is the humblest of slaves; his, to do with as he wills. A daughter of Belgravia, with Lady Beranger's worldly-wise notions dinned into her ears, and with worldly, ambitious women examples for her in daily life—of this man she wants nothing, only *himself*; to gain his love, and above all, to be let to love him, she would fling all other considerations to the four winds without a murmur or a regret.

In a sort of maze, she lifts up a pair of big, incredulous black eyes to him now—eyes so soft and wistful—so filled with newborn light that no one would believe they belonged to Gabrielle Beranger.

She forgets everything but him and the giant fact that he is hers. In spite of her peculiar nature and practical turn, she has pictured, like most of her sex, a paradise of love about this man, and lost in the golden vision of Love's paradise gained, she lets her usual scepticism slip out of her mind, and only knows that Lord Delaval, whom she has worshipped for three years with the feverish fierceness of her Bedouin nature, *is* wooing her—strangely and abruptly, but in the sweetest, subtlest way that a man can woo. Gabrielle is sharp as a needle, yet it never crosses her brain in her lovesick frenzy that *real* feeling is *not* eloquent in expression, and that when a man *really* craves anything and trembles lest he should not grasp it, flowers of rhetoric are usually denied to his tongue.

She sits spellbound, with drooping lids. Literally *nothing* seems to live in her, save a vivid sense of his words, and the intensity of their meaning. Her keen intelligence is lulled to sleep, her habit of doubting is dead, *pro tem.* She does not try to subject his protestations to any analytical process; they only seem to float through her mind in a kind of soft mist, and she sits white now and silent, and feeling, as she thinks she can never feel again, content, almost in a dream, and yet full, awfully full, of an intensified vitality.

"I want to tell you, Gabrielle," Lord Delaval says very low, while his audacious arm steals round her magnificent shoulders and her crimson cheek is pillowed on his breast, "that I love you as no one has ever loved you, and that I am *determined* to win from you all that I wish! I have *never* been baulked yet, if I determined to reach anything. If I preserve my will intact, I shall not accept anything but the *whole* from you, the *whole*, sweetheart—do you hear? Of your heart and soul and body I will have all—*all!* or die unsatisfied. My hope to gain all this is by knowledge of your nature. It is you—*you* that I love, not a part of you, not an ideal being of you, not what you represent to other men's eyes, but what you are with your thousand imperfections, even blots. Nothing, Gabrielle, will change me towards you, for I have only given you what is yours by the law of affinity, and you, Gabrielle—well, I *defy* you to say that you are not wholly and solely *mine*."

It is masterful wooing this, insolent in fact, and it would revolt most women. Zai and even Baby, with her fast proclivities, would not understand it, and it would jar on their thoroughbred natures, but Gabrielle likes it.

The whole thing fascinates her—a visible shiver runs over her. Lord Delaval feels the shiver, and his arm draws her more closely to him, while the ghost of a cynical smile crosses his mouth. He stoops his head and looks full into her eyes, and then his lips rest upon hers, long and passionately, while her heart beats as wildly as a bird in the grasp of a fowler.

Luckily for her she *has* been partially imbued with a respect for Lady Beranger's beloved *convenances* and *bienséances*. Luckily for her, Belgravian morals, though they may be lax, are too worldly-wise not to know a limit.

Even while Lord Delaval's kiss lingers on her mouth she pulls herself away from him, angry with herself that she has allowed that long passionate caress, and yet feeling that she would have been more than mortal if she had resisted it. But she resolves to sift him, *au fond*, to find out at once if in truth the man is only laughing at her or whether, oh blessed thought, she has caught his errant fancy or "love" as she calls it.

"Lord Delaval!" she says, in a voice in which pride and shame mingle strangely together, "because I am a woman, with a woman's weak nature, do you believe me to be a fool? Do you think for a moment I deceive myself or let your words deceive me? Only last night you flirted horribly with Zai. Before, it was in Baby's ear you whispered your soft nothings. It was Baby's hand I have seen you furtively clasp. I know therefore that the love you profess for me is all stuff and nonsense! that playing with women's feelings is delicious food for your vanity. But *why* you should pick me out, why *I* should be a butt for you, I am sure I can't guess! I don't care to believe that because I am what Lady Beranger thinks me, that *you* want to insult me!"

A look of pain crosses her brow, and an appeal for forbearance, dumb but very taking, goes up from her eyes. Lord Delaval seizes her hands and holds them fast while his gaze bears steadily down on her.

"You should not doubt, Gabrielle! I have told you the truth, upon my soul! No woman's face can tempt me from you now. Whatever the past may have been, I swear I belong to you now and for ever! While I wait to claim you as my wife before the world, and I *must* wait, for reasons which will be satisfactory when I tell you them, you will go on doing as you do, draining men dry to the one drop of their souls that you can assimilate. But that is not love, though they may lay their lives and fortunes at your feet. Aylmer would never satisfy your heart, Gabrielle, but you may flirt with him if you like, and drive him mad by these sweet eyes, these soft red lips," and he lifts up her face and studies it for a moment, "so long as when *I* want you, you come to me at once. It will be no sacrifice on your part, for you will only be obeying the law of your nature in loving me and I—I shall take you not as a gift, but as a right, my Gabrielle!"

Before she can answer him, he has taken her into his arms, and rained down kisses on her brow and cheeks and lips and is gone, with the conviction in his mind that, if he wishes it at any time, it will not require much pressing on his part to mould *this* girl's future to his will.

True he does not care a snap of his fingers for her, but any woman, beautiful of face and form, is not an object to be disdained or rejected, and Lord Delaval is not the only voluptuary among the Upper Ten.

Alone with the gathering shadows, and still wrapped in the presence that has left her, Gabrielle sits for an hour undisturbed. In the latter days she has thought several times that Lord Delaval had begun to recognise her claims to admiration, in spite of his flirtations with Baby and Zai, and alas! for Belgravian nurturing, it is a truth that the consciousness that her attraction for the man is *only a physical one*, in which her brains and soul bear no perceptible part, is far from being an unpleasant sensation.

"How very shocking!" a few prim spinsters may exclaim, but it is nevertheless the truth and nothing but the truth. It may be that most women love to conquer with the legitimate weapon, *beauty*, of the sex.

Poor plain Madame de Staël would willingly have exchanged all the laurels men laid at her feet for the tiniest, meanest blossom offered in a spirit of "*love*" or "passion" by them to women whom she justly regarded as her inferiors.

Gabrielle forgets her cross, her mother's low birth, Lady Beranger's taunts and everything else unpleasant, as she positively revels in a sense of Lord Delaval's admiration.

Rising from the lounge, she walks to the mantelpiece, and placing her elbows on it stares in a fixed, almost fierce way, into the mirror.

The shadows that flit over the room are broken here and there by a few last dying sunbeams, and her beauty is improved by the flickering light. The sweet eyes and soft red lips to which he had alluded, gain fresh merit since they are decoys to his erratic fancy, and have fanned the spark she has tried to ignite into a flame that has at last burst into words.

Then between her and the mirror the superb face of her lover rises up, and the cheek that has just been pressed against his breast glows a lovely carmine, that is wasted on the unappreciative dusk, as she clenches her little fist, and swears in true and forcible Bohemian fashion to bring all her woman's wit to aid in winning this man for her husband.

Just at this moment Lady Beranger walks in, and without noticing her stepdaughter by word or look, throws herself a little wearily into an arm-chair.

"What are you thinking of, *belle mere*?" Gabrielle asks after a little.

"Thinking of! There is plenty to think of I am sure," Lady Beranger retorts curtly. "I shall never be at rest till the girls are safely off my hands; unmarried daughters are the greatest responsibility breathing."

"I will try and lessen your burden," Gabrielle says, in a bland voice, but with a curl of her lip which the dusk hides, "I'll promise not to say 'no' if anyone asks me to marry him."

Lady Beranger laughs a sharp unpleasant laugh.

"It is not likely *you* will lessen my burden!" she says sharply. "Everard Aylmer, who was my forlorn hope for you, told me he was off directly for a tour in India, so *he* is not going to ask you."

"May be, but then you see, there are other fools beside Sir Everard Aylmer, in this world, Lady Beranger," Gabrielle answers flippantly, as she saunters out of the room.

"Hateful girl!"

And having relieved herself of this, Lady Beranger settles herself more comfortably, and begins to build castles in which Zai and Lord Delaval, Trixy and the fascinating Stubbs, and Baby with her elderly *inamorato* figure.

"That actor fellow showed his cards well last night," she soliloquises. "He is after the Meredyth filthy lucre of course, so now there's every chance of Zai catching Delaval. Trixy is thrown away on that dreadful cub, but after all, it doesn't much matter who one marries. After a month or so, now-a-days, the women think twice as much of other people's husbands as of their own. Baby will be all right in Archibald Hamilton's keeping. That child really frightens me by her defiance of everything, and I shall be truly thankful to wash my hands of her before she goes to the furthest end of her tether. As for Gabrielle," a frown puckers her ladyship's patrician brow, "I wonder who she has got running in her head? I *hope* it is not Delaval; a neck to neck race between her and Zai would end in her winning by several lengths. Zai, though she is my own child, is the biggest little fool, with the primitive notions of the year One, and I *can't* alter her, worse luck!"

# CHAPTER III
# "FROGGY WOULD A WOOING GO."

"Gold, gold, gold, gold,
Bright and yellow, hard and cold;
Molten, graven, hammered and roll'd,
Heavy to get, and light to hold,
Price of many a crime untold."

"POOR Mr. Stubbs," sneers Gabrielle.

"Poor Mr. Stubbs," says Zai.

"Poor Mr. Stubbs," laughs Baby.

And with very good reason.

It is his eighth visit.

Trixy has deserted her downy nest among her cerulean cushions, and sits bolt upright on a tall-backed chair. To-day is devoted by her to the personification of "Mary Anderson."

Her attire is of virgin white, not flowing in undulating waves of Indian muslin, or ornamented by tucks *à l'enfant*, but falling in severe satin-like folds round her beautifully moulded figure; her wealth of yellow hair is gathered at the back of her dainty head in a classical knot, traversed by a long gold arrow. She wears no bracelets or rings to mar the perfect whiteness of her arm and fingers, and while one hand toys lazily with a mother o' pearl paper-knife, the other rests on a well-thumbed copy of "The Lady of Lyons."

Opposite her, but at a discreet distance, her Claude perches nervously on the edge of his chair; his face has acquired more flesh and blood with his increased importance as the *fiancé* of the beautiful Miss Beranger, and his puffy cheeks glow like holly-berries under her glance.

Not that her glance by any means shows the odalisque softness, of which mention has been made; on the contrary, there is an incipient loathing in it, that she tries to conceal under the shelter of her long golden lashes.

But everything nearly has two sides, and the white drooping lids find favour in her adorer's sight, for he attributes them to the delicate shyness peculiar to the *china* beings of the Upper Ten, and unknown to the coarse delf of his own class.

Once, and once only, has he ventured to lift the lissom white fingers to his hungry lips very respectfully, *bien entendu.*

It was the day when, Lady Beranger standing by, Trixy agreed to barter her youth and beauty for:

> "Gold, gold, gold, gold,
> Bright and yellow, hard and cold;
> Molten, graven, hammered and roll'd,
> Heavy to get, and light to hold,
> Price of many a crime untold."

BUT she had drawn back her fingers before they arrived at his desired goal, with a sudden hauteur that almost petrified him into a stone.

It was the first time he had been thrown in such close contact with "high life," and when it bristled up in aggrieved delicacy it appalled him; but the next moment, he awoke to a profound admiration for the maidenly reserve that was, of course, part and parcel of a refined nature.

Poor Mr. Stubbs! well may the Beranger girls pity him. He little dreams of the melting glances Trixy's sweet blue eyes have given to Carlton Conway, or how eagerly the hand like a snowdrift has gone out to nestle in Carlton Conway's clasp, and how the faint blush rose on her cheek has deepened into damask bloom when in the old days Carlton Conway whispered in her ear, nor how, tell it not in Gath! her pretty mouth had even pouted for Carlton Conway's caress.

But we all know that where ignorance is bliss, etc., etc. Ever since Mr. Stubbs has been duly installed in the dignified position of "future," to Lady Beranger's eldest daughter, he makes periodical visits to Belgrave Square.

As it has been told, to day is his eighth visit, but he approaches no whit nearer to his divinity as regards heart—in fact he has decidedly made a retrograde movement in her opinion.

Trixy fully realises the truth of the old saw, "distance lends enchantment to the view," and the nearer she sees him the more difficult it seems to her to swallow this big bitter pill, although it is heavily gilded. Still, she is determined to marry him somehow, for as regards more substantial things their hearts and such obsolete absurdities—she has fully realised the advantages and benefits this horrible sacrifice of herself, as she styles it, is likely to bestow.

What daughter of Belgravia hesitates long between love and ambition? That is, if she has been properly brought up? and how often are the marriages solemnised at St. George's or St. Peter's—marriages *du cœur*? A popular author writes of modern love—

"Though Cupid may seek for sweet faces,
From ugliness fly as a curse,
May sacrifice much for the Graces,
He'll sacrifice more for the—*purse.*
The priest, if inclined for truth's rigour,
Might write on each conjugal docket,
'When a lover's in love with the figure,
The figure must be in—the *pocket*!' "

And he is very nearly right.

Trixy has on a table that stands beside her two open morocco cases. In one, a magnificent necklet of diamonds sparkles and scintillates in the daylight, flashing back glances at a set of pigeon-blood hued rubies that repose alongside.

When her eyes rest on these the odalisque softness steals back to her limpid glance.

"Do you approve of the ornaments?" the millionaire asks nervously of his "liege ladye." He would not have ventured to say "Do you like me?" for all the world.

He is brimming over with gratification at his sumptuous gift being accepted, although Trixy has not had the grace to say even "thank you."

But then she is so sure of him that she does not trouble about common politeness.

"I have not yet learnt your exact taste, you know," he mumbles a little sheepishly, reddening to the roots of his more than auburn hair, possibly with the pleasurable vision of the time when he *will* know Trixy's taste better.

*Poor* Mr. Stubbs!

At present she is still "doing" Mary Anderson, and may be a statue of Galatea for aught he can find in her of warmth, or learn of her tastes and feelings.

"The ornaments are very well," answers this often-to-be-met-with type of Belgravian daughters, with an insolent indifference which is quite assumed, for such costly baubles are her heart's delight. "I should certainly have preferred sapphires to rubies. They suit blondes so very much better."

Poor Mr. Stubbs feels and looks extremely disappointed, and crestfallen. He has paid such a very large sum for the rubies. He has ransacked all the leading jewellers' shops that the stones may be large, and flawless, and the exact colour of pigeon's blood, and here is his reward.

For a moment it seems to him that there is something a little disheartening and depressing in aristocratic coldness and ingratitude, and that some of the gushing thanks of little Imogene of the Vivacity, or pretty Vi Decameron of the Can-Can Theatre would not be amiss, but only for one moment does his tuft-hunting soul turn traitor to the high life it adores, and he quickly brightens up.

"If you will allow me, I will take back the rubies, and desire sapphires to be sent instead."

"Oh, no, no! it would scarcely be worth the trouble of changing them, these will do very well," she answers in a tone of languor, but she remembers the vulgar old adage of "a bird in hand is worth two in the bush," and to put a bar on any chance of losing the disparaged rubies, she quietly clasps the morocco cases, and locks them into an ivory and ebony Indian box.

The big drawing-room in Belgrave Square is very dull. From outside comes the rush of vehicles, and the June sunshine tries to peer through the closed jalousies that fine ladies love. The clock ticks rather obtrusively, but Trixy likes to hear it, for it tells of the flight of time; a prospect she has at heart at this moment, and a short silence falls upon as ill-assorted a pair as ever a longing for the world's vanities has brought together. Looking at them, the story of Beauty and the Beast presents itself, excepting that the Beast is not likely to turn out anything else, save as far as riches are concerned. From the day Mr. Stubbs popped the question, as Baby has it, and Trixy accepted him, Lady Beranger has thankfully thrown off the onus of chaperonage, which a rigid adherence to her beloved *convenances* insisted on before, and long *tête-à-têtes* are vouchsafed to the "happy young couple," as she calls them (Extract from the Stubbs' family bible—Peter Robinson Stubbs, born July 12th, 1820, rather upsets the word young), but her ladyship cannot stand the man in spite of his youth and happiness, and slips out of the way whenever his loud knock resounds through the mansion. She has no fear that Trixy will prove refractory now that the die is cast, and the match has been announced formally in the columns of the Court Journal and other Society papers. Besides, a dissolution of the contract would involve a return of very expensive presents, including the despised rubies, and Lady Beranger's insight into human nature, or rather into her eldest daughter's nature, leads her to think rightly. Trixy is her mother's child to the backbone.

In spite of her utter loathing for the man to whom she is going to swear glibly love and eternal fealty, she has received too heavy substantial tokens of his regard to allow her golden calf to drift away. She has thoroughly made up her mind—such as it is—to cast away all romantic nonsense, *i.e.*, her adoration of Carlton Conway, for the sake of worldly benefits, and now it is ten to one, that if the all-conquering C. C. came in his noble person to woo her, she would deliberately weigh against his undeniable fascination the prospect of being a leader of Society, with magnificent diggings in Park Lane, and the very comfortable sensation of a heavy balance at Coutts'.

"You think you really would prefer Park Lane to Carlton Gardens?" Mr. Stubbs inquires deferentially.

Under the powerful glamour of Trixy's beauty he feels as if he could buy up the Fiji Isles, or even that very uncomfortable residence, Bulgaria, if she wills it. Of course, she likes the big house in Park Lane. What woman, especially a daughter of Belgravia, would not? with its superb array of balconies, and galleries, and conservatories, and its vast reception rooms, where Trixy fully intends to queen it over other leaders of Society, but she just bends her pretty little yellow-crowned head in assent.

One may have a dancing bear, but one is not forced to converse with him, she thinks, and she gives him a long, level look, wondering what animal he is really like. Gabrielle had likened him to a frog, but he is too bulky for that; a bear or a buffalo, she decides, and while she does so, he has come to the decision that no cage can be too gorgeous for his radiant Bird of Paradise, and he glances, but covertly, at her in a sort of maze at the curious freak of fortune that is going to bestow on him such a *rara avis*.

He looks sideways at her sweet scarlet lips, and marvels what he has ever done in his prosy money-making life to make him worthy of their being yielded to him—not yet, no, *certainly*—not yet, he is aware of that, but perhaps, some day! He gloats with an elderly gentleman's gloating on the supple young form and perfect face, and quite a delightful awe creeps over him at the very idea of the future presence of this flesh and blood divinity at his hearth and board.

Nature has not been munificent to him in the way of looks. He has a broad, florid, rather flaccid physiognomy, and his proportions are not symmetrical, but taking him all round, he is not a bad sort, and he has a good heart.

True it beats beneath a huge mountain of flesh, but, never mind, it beats all the same with a good deal of honest warmth. His feelings towards his fair autocrat are a mixture of profound admiration and profound gratitude—the last sentiment being born of the first.

Gratitude is in fact an intensely tame word to express what he feels for Trixy's munificent gift to him of herself. With all these feelings rife in his very broad breast, feelings that would gush forth eloquently in most men, Mr. Stubbs remains strictly practical and common-place, and fortunately his wife elect is better able to sympathise with him as he is than if Cupid spoke from his lips in flowers of rhetoric.

"And the furniture? From Jackson and Graham's, I suppose?" he asks deprecatingly, as if it was *her* money and not his that was to pay for it.

"From Jackson and Graham's of course! You surely are not thinking of going to Tottenham Court Road, Mr. Stubbs?" Trixy says raspily, with a little sniff of her Greek nose.

"No, no! *of course* not!" he murmurs alarmed.

"Remember, I cannot have any hangings but *blue*—blue suits my complexion, you know; not *dark* blue, mind, but *bleu de ciel*!"

"Blue, certainly," he answers humbly, much more humbly probably than Jackson and Graham's foreman would.

"And Mr. Stubbs, pray don't forget that I hate anything modern. I like everything *old*, in *furniture* I mean!" she says, warming up with her subject. "Chippendale and all that sort of thing."

"Florid carving you would like of course?"

"Florid! Horrid! Plain chairs, with shields at the back for the—— "

She stops suddenly, while a look of disappointment and dismay creeps over her face.

"But you haven't a crest, have you?" she adds, with as much solemnity as if she were asking "Have you hopes of salvation?"

"A crest?—of course I have!" he replies jauntily, not a bit offended at her doubt on the subject. "A sweet little crest. It has a little turretted house on the top, with what they call in heraldry a martinet perched on it. I don't understand much about birds, but in plain English, I expect it's a swallow, or maybe a tom-tit. And the motto is a very nice one, and very applicable too—*Fortes fortuna juvat*," and he smiles complacently.

Trixy has a horrible suspicion that he also winks.

"I don't understand Latin," she says scornfully. "You see, they don't teach it at fashionable schools. It is a language that does very well for prescriptions and things, and is only fit for doctors."

"I know a little Latin, and my motto in English is 'Fortune favours the brave!' " he explains pleasantly, with another affable smile and meaning look, which are quite lost on Trixy, whose worst enemies cannot accuse her of any undue 'cuteness, as the Yankees have it. She has no more idea that the man is alluding to himself and herself than if he was speaking Greek, which is another of the languages she knows nothing of.

The only thing that strikes her is how funny he would look if his bravery was called into account, and how slowly his short stout legs would carry him, if he ever wanted to run away from an enemy.

"You say the crest has a castle with a bird on it. That will do I fancy on the furniture. People don't trouble much about the subject, so long as there *is* a crest to make the things look more aristocratic. Can't the Beranger motto be added to yours? It is French, and everybody knows French."

"May I ask what it is?" he asks wondering how he can have overlooked it in his diligent researches into "Lodge" and "Burke " and "De Brett," works that, bound in velvet and gold, have prominent positions in his library.

"It is '*Noblesse oblige*,' 'Nobility forces,' you know."

Mr. Stubbs reddens as he thinks the addition she suggests will very likely provoke a smile from ill-natured people, who *might* fancy that the Hon. Trixy Beranger's finances forced her to become the Hon. Mrs. Stubbs.

"I don't see how it can be done," he remarks. "It would be going against the rules of heraldry I am afraid."

"What does *that* matter?" she cries captiously. "It would be very hard if I really set my heart on anything, to be done out of it just because some stupid sign-painter's ideas did not coincide with mine."

"Heraldry is not exactly sign-painting, it is a science," he ventures to remonstrate, anxious to smooth down her ruffled feathers.

"Really, Mr. Stubbs, you seem to think my education has been dreadfully neglected! I was five years at Mrs. Washington de Montmorency's *élite* establishment for daughters of the nobility only! Then I was at Madame Thalia de Lydekerke Beaudesert's finishing academy for *la crême de la crême* only, and Lord and Lady Beranger have spared no expense in educating me! Signor il Conte Almaviva taught me Italian, Rubenstein considers me his show pupil, Patti was heard to say that she envied me my voice, and—and—of course I know that heraldry is a science, but science or no science, I cannot see why I should not have exactly what I want carved on the backs of my own chairs and sofas. However, it really isn't worth the trouble of discussing," and Trixy half-closes her eyes and falls into languor, a manner beneath which he invariably feels the social gulf widen between them.

He cannot, even if he tries, affect this supreme indifference, this delightful repose that sits so easily on Lady Beranger and her belongings.

Leaning back against the *Prie Dieu* chair, with half-closed eyes, Trixy looks like a marble effigy of Resignation, but she does not show the gentleness and patience with which the virtue of resignation is generally invested. She is rather a cold, hard martyr to untoward circumstances, with a big wall of ice raised up around her that seems to freeze up her companion.

Surreptitiously he glances at a monster watch, like a bed-warmer, with half-a-dozen gaudy seals and charms attached to it. He really is anxious to find that the three-quarters of an hour, which Lady Beranger had hinted to him was the proper term of a courtship, are up; but time has not flown on the wings of love, there are yet ten minutes wanting, so he settles himself in his seat, and just escapes the sight of Trixy's pretty mouth elongated in a long yawn.

He commences a sort of auctioneer's catalogue of the worldly goods and chattels she will possess directly she is mistress of Park Lane, divining that this is a subject which really interests her, and hoping to make her forget about the crests and mottoes.

Thoroughly mercenary himself, he quite understands how pleasant it must be for her to know all she will gain as his wife. Exchange and barter are household words to him. Ever since he was in knickerbockers and short pants he has been buying and selling, and he sees nothing at all extraordinary or revolting in this young person giving him her youth and beauty in exchange for his money.

Love! Well, love to his fancy is an excellent thing for boys and girls, but Mr. Stubbs has reached an age when passion *ought* to lose most of its fierceness and glamour, and a placid liking sound more comfortable.

He has given up business now, so he knows he will be usually at hand to guard his beautiful wife from the impudent swells—idle, good-for-nothing specimens of the *genus homo*—to whom morality is an unknown word, and whom he dislikes thoroughly, though he is deferential to their faces.

So that on the whole his matrimonial scheme bears a remarkably smooth aspect.

"There are one or two other little things on which I should like your opinion before I write my directions."

Hearing which she brightens up at once into an attitude of interest.

"Did'nt you say the other day that you preferred a brougham to a clarence?"

"A brougham by all means, and it *must* be by Peters."

"Have you a particular fancy for Peters?"

"Yes, yes. He is the only maker who is *chic*. Most of the others turn out heavy lumbering vehicles, with not the style about them that would suit *me*; but then you see, we have always been considered to be so very *difficile* in our tastes, and the brougham *must* be green."

"With scarlet under carriage, and body well picked out with broad scarlet lines?"

"No, no! Picked out with black," she says very decidedly, wondering at the awful taste of the man. And there is not a doubt but that his taste *is* showy, he wears at this identical moment a miniature yacht in full sail, in gold and enamel, as a scarf pin, and a tie of violet satin, with orange stripes; orange is in fact his pet colour, from rhubarb down to the primrose of his gloves.

"Yes," she says, as if reflecting deeply, "the brougham must be green, a *very* dark green, and picked out with black, and brass mountings."

"A little sombre, don't you think?" he suggests timidly.

"Good heavens, Mr. Stubbs! Do you want me to drive out only on the ninth of November and look as if I was a part of the Lord Mayor's show?" she asks excitedly, raising her voice and causing him to give a little jump on his chair.

It is the first time she has displayed any variation of feeling, and the spice of devilry in her eyes, though it does away with Mary Anderson, heightens her beauty. Usually Trixy Beranger resembles a large waxen doll, with yellow hair and pink and white cheeks.

But she recovers her temper directly. It strikes her that this glittering fish may prove a slippery one if she allows the stormy side of her character to burst out before the matrimonial noose is tied.

"But, of course, I know you were only joking about the colours for the brougham. I am *sure* your taste is similar to my own, and that you think nothing can be too quiet to be aristocratic. Mamma rather wants me at four o'clock, have you any idea what the time is?"

He glances once more at the leviathan timekeeper he carries, and discovers that he has outstayed his limit fifteen minutes, and that his regular constitutional before feeding time will have to be curtailed.

"I, too, have numerous letters to write, so I think I'll say *au revoir*."

Trixy sticks out five fingers carelessly, and he takes them in silence, but he is not bold enough to squeeze them ever so little, and he breathes more freely directly he is outside the big drawing-room door.

His broad back turned, Trixy steals out on tip-toe upon the landing, and when he is fairly out of the house, she opens the ivory and ebony box, takes out the two morocco cases, and walking up to the large mirror opposite, she leisurely puts the chain of brilliants and the band of rubies round her snowy throat. Rubies flash in her ears, and a huge bracelet of the same gems gleams blood-red on her rounded arm.

For a minute or two she gazes enraptured at herself, then she rushes up the stairs, two steps at a time, like a tomboy, and bursts like a whirlwind into what is called Baby's school-room.

Baby has for some time given up instructive books for more refreshing waters of literature in the shape of French romances; but she still clings, with the small amount of tenacity there is in her nature, to the old ink-stained table and hard chairs, in whose company she tottled up four and four, and invariably made them nine, and wept bitter tears over the dry food provided for her mind by Miss Jenkinson, a staid sanctimonious old spinster that Lady Beranger had picked up out of the *Guardian*, and who, for twenty pounds a year and her laundress, agreed to the herculean task of bringing up the youngest Miss Beranger in the way she should go, so that when she was old she would not depart from it.

Alas! Miss Jenkinson's counsels have fallen on stony ground, for Baby is the biggest young reprobate that ever danced through life in kittenish glee and kittenish mischief.

The school-room, now that Miss Jenkinson is gone, probably through worry, to a premature grave, is used as a sort of *omnium gatherum* for all the Miss Berangers, and here they gather usually when not *en toilette* and *en evidence.*

"Look at me," cries Trixy in a shrill voice, "and admire me."

And jumping on to the centre of the table she stands with a half-conscious, half-comical expression on her face that elicits a burst of laughter from the other three.

"How *can* old Stubbs make such a fool of himself? He *must* know you are only marrying him for those things!" Gabrielle says contemptuously.

Trixy takes no notice. Gabrielle is not a pet or a pal of hers, and Gabrielle's wits are too sharp for her.

"I say, Zai, what *wouldn't* you give for such beauties as these?"

"Nothing! I don't care a bit for jewels, and I wouldn't accept such costly gifts from a man I did not care about for anything," Zai answers quietly, going on with her drawing.

"Grapes are sour, my lass. The man you *did* care for might not be able to give you them," Trixy says spitefully.

"*I* would accept them fast enough if I had the chance," Baby confesses ruefully, climbing on to the table as well, and enviously examining the brilliants and rubies. "Just fancy, that old Hamilton has never offered a thing but that!" and she sticks out her third finger, on which reposes an old-fashioned ring, with a bit of Archibald Hamilton's sandy hair shining through the crystal. "Scotch are such screws, I hate them. Do you know, girls, that I have nearly made up my mind to give the old gentleman the slip, and to elope with Gladstone Beaconsfield Hargreaves."

"Heavens! what a name for a common village Veterinary," Gabrielle says, with a curl of her scarlet lip. "And to think of his awful people having the audacity to mention Beaconsfield in the same breath with *Gladstone*!"

"Rather mentioning Gladstone in the same breath as *Beaconsfield*!" cries Zai, horror-struck. She is a thorough little Conservative to the back-bone, and even goes to sleep in her dainty white-curtained bed with a badge of the Primrose League upon her bosom.

"A very good name it is!" flashes Baby, taking up the cudgels in defence of her rustic admirer. "I think his godfather and godmother were sensible people, and had no narrow-minded party-feeling and that sort of rubbish in their heads. Real Liberal-Conservatives they were, of course. I can't stand politics, Trixy, can you?"

"Can't abide them," Trixy murmurs lazily. "I hate everything it gives one trouble to understand."

"Politics make me quite ill," Baby goes on, as she jumps off the table and flings herself full-length on the hearth-rug. "When the governor and

Lord Delaval begin at them, I always feel inclined to roar. The governor shuts up one eye, and tries to look so awfully clever, you know.

" 'Dolly Churchill, my dear fellow, is the man—*the man*! Our *only* hope in these days of misguided, dangerous democrats. Our *only* stay! The Liberal Government have been the very devil—they have played ducks and drakes with everybody and everything, and if they had lasted one day longer—*one day longer! mark my words!*—we should have been at—at—well, *not where we are now*!'

"And Delaval, who is a red-hot Republican at heart, just smiles that beautiful cynical smile of his, and thinks the governor a regular jackass, and so *do I*."

"You shouldn't speak so of Papa, you irreverent monkey," Zai says gravely.

"Shouldn't I *really*!" Baby replies, mimicking her voice. "Well, then, I *will*. I love my Papsey. He is a dear old boy, but all the same, I don't think he will ever set the Thames on fire with his brilliancy. Why, ever since he has been in the House he has never said anything but 'hear, hear!' or joined in the ironical cheers."

"Lord Salisbury thinks a lot of the governor. I heard him say to Count Karoly the other night that Beranger was one of the most reliable men in the House, and so very cautious," Zai says quietly.

"No wonder, as he never opens his mouth," Baby laughs. "What do they have a lot of dummies for in Parliament?"

"Oh, just to make the whole thing look more imposing than it is, I suppose," Trixy drawls languidly. "Very likely they prefer most of the members not speaking, as the stupid ones might let out the secrets to the Opposition."

"*Gladstone speaks!*" Gabrielle announces solemnly, as if it is not a remarkably well-known fact. "He has been known to speak for three days and three nights without pausing to take breath even, and his eloquence has so overwhelmed the House—— "

"With sleep, that no one ever got at the real meaning of his speeches," interrupts incorrigible Baby. "Any way, the Irish didn't. My Hargreaves is an Irishman (that is why he was christened Gladstone Beaconsfield I dare say. The Irish muddle up politics so, you know), and he told me that in Paddy land *Gladstone* is the new name for *Blarney-stone*."

"I wish you would not regale us with the imbecile witticisms of your Vet, Mirabelle," Gabrielle mutters crossly, for she worships the G.O.M., and feels a slash at him acutely. And Baby knows she is wroth, for it is in ire only that she calls her Mirabelle, but Baby cares for nothing or nobody.

"My Hargreaves is *not* a vet, now. He is assistant riding-master to the great Challen."

"Baby, is this why you coaxed the governor into letting you have riding-lessons?" Zai questions anxiously.

Baby springs up from the hearth-rug, and turning a pirouette, pauses beside her pet sister.

Leaning over she whispers in her ear:

"It is, but if you promise not to peach, Zai, I'll tell you something about—— "

"Who?" Zai whispers back, colouring vividly.

"C. C., but not before Gabrielle and Trixy."

Zai blushes more deeply still as she bends over her drawing, and wonders if the letters C. C. will always send the blood surging over her face and set her pulses throbbing.

In spite of his heartless conduct at Elm Lodge she loves him dearly still, and lives from day to day in the hope that the clouds will clear away, and give her back the sunshine of life—Carl's love and presence.

And as she sits and drops off into a sweet waking dream, Gabrielle's voice startles her, and drags her back into everyday existence.

"Seven o'clock! We must be off and dress for dinner. There goes the first bell. Zai, there's a treat in store for you to-night."

Zai looks up, the dreamy expression still lingering in her eyes. A treat! For one moment she really fancies "he" is going to appear somewhere or somehow, but the next instant she fully awakens to her folly.

"Lord Delaval dines with us to-night, and afterwards we are all going to the theatre."

"*What* theatre?" Zai asks quickly.

"The Bagatelle, to see 'Hearts *versus* Diamonds.' "

"And '*him*!' " Zai thinks to herself, waxing white as a lily at such an ordeal with Lord Delaval's mocking smile before her, and Lord Delaval's cold, keen gaze watching her face.

"Who sent the box for to-night?" she asks, for she knows Lady Beranger never spends her money on such things.

"Lord Delaval."

Zai colours again, and stoops down on pretence of picking up her pencil. She *feels* that Gabrielle is looking at her.

"That man has sent it on purpose to vex me," she thinks. "I detest him."

# CHAPTER IV
# AT THE BAGATELLE THEATRE.

"Why did she love him? Curious fool, be still,
Is human love the growth of human will?"

WHEN Lady Beranger and her party enter a large stage-box and settle themselves noiselessly in their seats, the first act of 'Hearts *versus* Diamonds' has begun, and the big bass is booming out a lugubrious overture to Ferdinand—the deserted lover's reproaches to his faithless and diamond-worshipping Lady Yolande.

On the whole Carlton Conway looks superbly handsome and effective, when, as Ferdinand, he takes up a highly picturesque pose right in the centre of the stage. His head erect, his chest well thrown out, a little after Kyrle Bellew; his shirt-front ample; his tail-coat, and waistcoat and trousers, his patent leather boots, unimpeachable; and a gardenia from Hooper's, in Oxford Street, although he can ill-afford the half-a-crown paid for it, fresh and snowy and fragrant, reposing on his broad breast.

With one white hand uplifted, the forefinger pointing in scorn; the third finger sparkling with a tiny but pure brilliant (Zai's gift), he hurls:

"Oh, cursed hunger of pernicious gold,
What bands of faith can impious lucre hold?"

in a deep, impassioned voice, that fairly electrifies his audience, but makes very little impression apparently on the Lady Yolande, who has quite made up her mind to give up love and poverty for a comfortable mansion in Mayfair and plenty of diamonds and money.

Miss Flora Fitzallan, as the Lady Yolande, is at her best to-night. She looks, in fact, as if a whole page of "Debrett" was devoted to her ancestry, thereby proving that we are not what we seem, and often seem what we are not.

In the palest of blue brocades, heavily embroidered with silver, and a tuft of pale blue ostrich tips placed jauntily a little on one side of her head, and a long Court train, edged with the very best imitation ermine, she looks quite good enough for a leader of Society.

On the finger of scorn being pointed at her, the Lady Yolande laughs tragically, and with an artistic twirl of her skirt swoops down close to the foot-lights, and while her glance roves over the *jeunesse dorée* gathered in the stalls, cries in a contralto voice:

My name is Blue-blood! In the House of Lords
My father sits and has his say;
My mother was a Mistress of the Robes,
Before those awful Tories had their sway!
Thou forgettest, Ferdinand, that sangre azul flows
Through all my veins; that in my face
Not only love, but high ambition glows,
With which, alas! thou never canst keep pace!
Lapped in soft luxury, born in marble halls,
Vassals and serfs to answer to my calls,
I could not brave the humiliating woe
Of in this world coming down so low.
Ferdinand, forgive me! and let me go!
Without my purse full, I should surely pine,
I love good dinners, and I love good wine;
My beauty decked in velvets, satins, lace,
A jewelled diadem to crown my face.
Ferdinand, I leave thee! heart-broken, with a sigh,
But without gold and diamonds I should die!—die!

UPON this confession Ferdinand shows the laceration of his feelings by striking another attitude, an attitude of giant but picturesque despair. He folds his arms tightly across his chest, strides heavily towards her, and wears generally a depressed appearance.

"Oh!" he exclaims, lifting up his fine eyes to the gods in the gallery. "Lend me, I pray, strength to bear her perfidy."

As his glance slowly travels earthwards he espies Zai, and starts slightly, but the sight of her sweet face gives real pathos and eloquence to his voice as he murmurs tenderly:

"Yolande! Beloved Yolande! Thou knowest not the vulture that gnaws my heart, or thou would'st pause in thy fiendish work. False Yolande! Thou hast *never* known what heart is, but—

" 'I will tell thee what it is to love.
It is to build with human thoughts a shrine,
Where Hope sits brooding like a beauteous dove,
Where life seems young and like a thing divine.
All tastes, all pleasures, all desires combine,
To consecrate this sanctuary of bliss.
Above, the stars in cloudless beauty shine.
Around, the streams their flowery margins kiss,
And if there's Heaven on earth—that Heaven is surely this!' "

CARL CONWAY is really a very fair actor, and his voice is both musical and *entrainante*, and he spouts these lines with a wonderful passion and softness that appeal to all the women present, and as he speaks them, ever and anon his handsome brown eyes rest a second on the stage-box where poor little Zai sits well back in her corner.

Her eyes fixed on the beloved face, she forgets the existence of anyone else, her cheeks are flushed with excitement, her heart throbs fast, and a suspicion of a tear shines on her long lashes. Not a word does she utter, not a word does she hear; engrossed in this, the first love of her life, the play itself goes on without her taking in the gist of it. All she sees is Carl—Carl, with his superb face, and with his eyes full of the old, old passion as they linger on her and seem loth to turn away.

The curtain falls and rises twice over, and she thanks Providence that for once her people leave her alone so that she may gaze her fill. Who knows when they two will meet again—and how?

The girl's poor heart grows cold as ice when the *dénouement* of the play comes, and Ferdinand, praying for the boon of a last kiss, the Lady Yolande yields her proud lips to him.

Yields them *con amore*, too, it seems to Zai, as she shrinks back from the sight with a jealous pang that makes her shiver and clasp her little hands desperately together.

Then the curtain falls for the last time, and she looks up and catches Lord Delaval's eye.

It seems to be searching her very soul with a fixed, keen gaze that has something regretful about it, though his lips have a half-mocking smile.

"That fellow, Conway, really acts tolerably," he says aloud to Gabrielle. "Did you notice the ring of pathos and truth in his voice? And yet those sort of chaps lead such a hollow life of shams and tricks, that they can't possibly have a genuine feeling in them. What do you think of Flora Fitzallan, Miss Beranger?"

"Just what one thinks of such creatures," Gabrielle answers contemptuously, "outside all paint and powder. Inside—— "

"Pray don't give your opinion on people like Miss Fitzallan, Gabrielle. They are not fit subjects for your discussion; at any rate before me and my daughters!" Lady Beranger remarks severely.

Gabrielle elevates her brows and shrugs her shoulders. Then, as her stepmother sweeps away, she says:

"I think one thing about Miss Fitzallan, Lord Delaval. I think she has a *grande passion* for Carl Conway, and I expect she does not try to hide it—*off the stage*!"

And Zai hearkens in bitterness of spirit, but does not love Carl one whit the less.

"I say, Zai, did you see that Lady Yolande kiss Carl? She kissed him *right on the mouth.* And I have heard that it is not *convenable* to do that sort of thing on the stage!" Baby whispers.

And still Zai holds her tongue, but as she listens, it seems to her that it is the last straw to break the camel's back.

# CHAPTER V
# CARYLLON HOUSE.

"You loved me, and you loved me not
A little, much, and over much;
Will you forget, as I forgot?
Let all dead things lie dead—such
Are not soft to touch."

FANCHETTE, having arrayed Trixy and Baby for the Duchess of Caryllon's fancy ball, finally seeks Zai. Zai—who still lies dreaming her love's young dream in the soft twilight, while a star or two peeps down inquisitively through the open window upon the increased loveliness that love has called up on her sweet face.

Regretfully she rises at Fanchette's entrance, and certainly no fairer daughter of Belgravia ever tripped through Belgravian *salons*. When her toilette is complete, Fanchette does wonders with her little artistic touches here and there, and Zai's costume, though simple, is exquisitely picturesque.

The bodice is long-waisted; the stomacher thickly embroidered in pearls; the Vandyke corsage is low in front, with a high ruffle behind, and the whole makes a beau-ideal of the old time Maestros; ropes of glistening pearls go round the slim throat and are wreathed in the chestnut hair. The dress of Blanche of Navarre is marvellously becoming, and would be becoming to a plain woman. What, then, must it be to this daughter of Belgravia, to whom Nature has been lavish in seductive tints?—this girl with a beauty so very fair that

"If to her share some human errors fall,
Look in her face and you'll forget them all,"

and who is very proud of herself, as she thinks that Carlton Conway will be at Caryllon House to-night, and will see how "nice" she looks.

Let us own that a woman must be composed of very strange materials who does not feel that it is charming to be young and pretty, considering that youth and beauty are the recognised weapons for slaughtering men's hearts.

Lady Beranger has always a fancy for "her own party" when she goes to a ball, and on this occasion the dinner in Belgrave Square has three additions to the family circle—Mr. Stubbs, Archibald Hamilton, and Percy Rayne—a connection of Lord Beranger's—a clerk in the Foreign Office, good-looking, harum-scarum, a pauper, and a detrimental. Lord Delaval was asked, of course, but had another engagement. When all her brood is gathered together, Lady Beranger, in silver *moire*, with the Beranger diamonds (but no! not the Beranger diamonds, for they are under safe lock and key and surveillance of one of the many Attenboroughs—but the duplicates in finished Parisian paste, which are quite as lovely and costly to the uninitiated eye), steps into the family landau.

They are late, and the crush of the room is uncomfortable beyond description, like all London crushes. But great as it is, Zai makes a decided sensation as she wades through the crowd on Percy Bayne's arm. Gabrielle is a Spanish gipsy; Trixy, Fair Rosamond; Baby, with her pink and white skin, golden hair, and white short draperies showered with rosebuds—a delicious piece of "Dresden"—but Zai to-night put every one into the shade. There is the usual quantum of sea-nymphs and flower-girls, characters from history and characters from fiction, of piquant costumes and of costumes which are chiefly remarkable for being *bizarre*.

As she and Percy Rayne fall into the line which just now is promenading the long room in the interludes of dancing, the Foreign Office clerk is conscious of that pleasant thrill of complacency—a sort of moral and even physical inflation—which a man feels when escorting a woman whose beauty glorifies her escort.

Zai's card is soon full—so full that only one waltz remains, which she guards pertinaciously. She is determined to valse it with Carl, even if the heavens fall. Several ask for it, but she laughingly says she is keeping it for a friend. That friend does not, however, seem in any haste to take advantage of her generosity.

She has been nearly an hour in the room before she even sees him, and then he is talking earnestly to Miss Crystal Meredyth, and only acknowledges *her* by a formal bow; and to add to this, Crystal Meredyth makes a very lovely Ondine to-night. How strange it seems to her that he should bow like this, when only a week or two before he looked at her with all his soul in his eyes, at the Bagatelle Theatre!

Zai's heart is full to bursting, and her red lips quiver a little; but while a weeping and gnashing of teeth is carried on inwardly, she returns his bow with one still more frigid.

And at this inopportune moment, Lord Delaval comes up to her.

"I think the next dance is mine?" he says, rather stiffly, offering his arm.

"You mistake," Zai answers.

She does not wish to go off with one man when she can stand here, the centre of a group of *jeunesse dorée*—all begging for "one turn," and this within earshot of Carl.

She would give anything to pique him now that he is so engrossed with this girl who has money.

"The next dance is Mr. Bayne's; at least his name is on my card," she goes on.

Lord Delaval bows—not a bow like the one Carlton Conway has given her just now, but a bow on the Grandison model. His taste and tact are perfect; nothing would induce him to dispute a point of this kind; but a look steals over his handsome face which is not common to it when Zai is its object—a look of cold hauteur, a look that has even a *soupçon* of dislike in it.

"I understood the dance was mine," he says, and quietly turning on his heel, he walks away. There are visible surprise and satisfaction among the butterfly youths at this little rebuff to the best match in Town—for lords of the creation, noble animals though they be, are yet creatures of weak mould.

But Zai's conscience smites her.

That the dance is Lord Delaval's she knew quite well when she allowed Percy Bayne to write his name over his. At the moment she felt a sort of perverse defiance of displeasure on the part of any man. But now she regrets having sullied her lips by a white lie, and she feels ashamed—as one always feels ashamed—when one has taken shabby advantage of the immunity which is chivalrously permitted a woman to do or say uncivil things by Society. It is a retributive justice perhaps, which accords her nothing for her incivility, for Carlton Conway, who is standing not far off, and alone—Miss Meredyth having gone off to dance—presently moves off too, without even a glance in her direction. It is really too much!

Blanche of Navarre's grey eyes sadly follow his retreating figure, and with a decidedly sinking heart, and forlorn spirit, she sees him a few moments after, careering "*au grand galop*" with his arm round Miss Meredyth's supple waist. Always that Miss Meredyth!

She feels wickedly vindictive against this girl—almost ghoulish, as though she would willingly scrunch her up, bones and all—this dollish beauty who has lured away her lover.

Zai grinds her to powder (mentally), under her high military heel, and turning to one of her adorers, asks for a pencil and deliberately writes down Lord Delaval's name for the dance she has reserved for Carl.

It is some time, however, before this tardy reparation becomes known. Lord Delaval feels that he has borne as much as aristocratic flesh and blood can stand from this girl, who seems so little aware of the magnificent distinction he has conferred upon her, and that it is full time to assert his dignity.

He asserts it therefore in the ordinary fashion of men who are *épris*—by bestowing his attention upon other women, of whom there are a multitude willing—and Gabrielle in particular—to accept everything or anything he chooses to offer, this Prince of Beauty, with his blond hair and ultramarine eyes.

Like so many poor boxes, they are ready to receive the smallest donation—a smile—a word—his arm for a promenade—or his hand for a dance. Yet even while apparently engrossed in wholesale flirtations with the fairest of the sex in the room, even while lavishing soft nothings, pressing fingers, he finds himself covertly looking again and again, and fervently admiring the slender figure in its old-fashioned quaint costume, the fair sweet face of the girl who he knows is over head and ears in love with "that actor fellow." Despite himself and his anger he cannot help secretly owning that never did woman exist more fitted to wear the purple, and to don the Delaval coronet than this one, and he resolves to win her—somehow.

Having "put down his foot" on this point, he feels that all flirtations with Carlton Conway, Rayne and all others must end, that he must clearly make it understood that such doings must stop.

Flirt though he has been himself ever since he dropped round jackets and donned the *toga virilis*, and flirt though he probably intends to remain until the very end of the chapter, he has not the slightest idea of allowing his wife to indulge in the same amusement.

No! no! no! a thousand times no!

The woman of his choice must be an exceptional being, and a very different stamp of woman to the puppets of the Belgravian *salons*, with whom he has been in the habit of dallying and associating, and with whom he has passed so many hours of agreeable foolery.

Cæsar himself may of course do what he likes, but we all know what is expected from Cæsar's wife.

It is an old, old story—carried down from generation to generation, and alas! for the honour of Society, a story infinitely more theoretical than practical.

The hours go on towards midnight—the crowd is suffocating, the heat intense, the gaiety at its height.

Since they entered the room, all the Beranger girls have been dancing, they are not the sort to personate wallflowers, none of them, and Zai in particular has not been five minutes under her mother's ample wing.

Instead of looking worn out, however, she seems in higher beauty and gayer spirits then usual, when Lord Delaval again approaches her.

"You are only just in time," she says, meeting his vexed eyes with a little laugh which he would think the most delicious in the world if he had not heard it bestowed upon any number of the golden youths during the last hour. "I have put your name down for this very waltz, and I was reflecting a moment ago whether I should have to send Percy to look you up, or whether I should give it to the multitude who are begging for it!"

Zai says all this with an air of delightful coquetry which is perfectly foreign to her. Poor child, she is of course only playing a part to hide her misery and mortification about Carl, but she plays it extremely well, and the coquettishness is remarkably becoming to her.

"I wonder you hesitated over the alternative, when there are so many to whom you could give the dance with satisfaction, no doubt, to both sides;" he answers a little sulkily.

"Yes! there *are* a good many," Zai admits with ingenuous frankness. "But, then, you see, I thought you really wanted it! If you don't—— "

"You know I do!" he cries, quite unable to resist the pure, soft, sweet face uplifted to him.

All his mighty vexation is scattered to the four winds as he looks down on her.

In this world everything repeats itself.

Like the judges of old—whose fiat was stayed by fair Phryne's face and form—so Zai's pretty grey eyes, snowlidded and blacklashed, and her smile, even though it be forced, disperse this man's anger in a trice.

As he speaks the band strikes up "Bitter Sweet," and putting his arm around her elaborately whaleboned waist, yet a dainty lissom waist in spite of whalebone, he whirls her away.

It is a glorious waltz—the room is lengthy, the floor well waxed, the lights glitter, and the music peals out an exhilarating strain, and these two have danced often enough together to know well the other's step and peculiarities.

It is also the end—though they don't know it—of butterfly flirtation.

A very fitting end, too, for flirtations.

In the end of some serious love affairs, so much faith and hope go down for ever that we might well play over them that *Marche Funèbre* of Chopin—that charming old Listz called the *Mélopée*, so funereal, so full of desolating woe.

But for the end of flirtations, what can, we ask, be more appropriate than the light, gay, and entrancing strains of the Bitter Sweet Waltz?

"You must be awfully tired! You had better let me take you somewhere to rest!" Lord Delaval says, rather tenderly. Zai is tired, and does not demur; and he takes her out of the ball-room into a long corridor, in which the waxlights are a little dim, and in which fewer flirting couples than usual are to be seen.

Like a huge maelstrom, the *salle de danse* has engulphed them, so there is not much difficulty in finding the quiet and secluded corner, free from interruption, of which Lord Delaval is in search.

He wheels a cosy velvet-cushioned chair near an open window, and when she has dropped into it he settles himself opposite her on the window sill.

Zai shuts her eyes, it may be from physical fatigue, or it may be that she does not care to meet the keen searching gaze—anyway, a short silence follows, during which she slowly fans herself, and he—well—he is considering how to plunge at once into the subject nearest his heart—for he hates to wait for anything.

"I don't care to talk about myself," he says, after a minute or two. "If there is an abomination in the world, it is an egotistical man; but I should like to know if you have ever heard things about me which have caused you to shun my society at times? I know I have a number of kind friends in Town ready to tell you that I am a flirt, and worship myself only."

"Yes," she answers, truthfully. "I have certainly heard your friends say both things of you."

"Perhaps in one thing they were right enough—I have flirted desperately in my life—every man who has never felt a strong exclusive attachment does flirt, you know, but never more! never more! I shall never flirt again—for—— "

He bends forward until his face almost touches hers, and whispers low—

"The strong exclusive attachment has come to me!"

Zai does not answer, though she flushes in spite of herself.

"You cannot doubt that I love you, Zai!" he pleads, passionately, "and that I shall be the happiest man on earth if I can persuade you to marry me. Zai, do you think you will ever care for me enough to do that?"

He catches hold of her hands, and holds them as in a vice, and though she draws them away, she does not rebuke him for calling her "Zai." Perhaps she scarcely heeds that he does so. She is sore at heart about Carl. She would give a good deal to show him that if he does not appreciate her there are others who do; and what could be a greater triumph for her than to leave the Duchess of Caryllon's ball the future Countess of Delaval. She would be more than the bright, gay, and rather spoilt girl Belgravia has made her if she did not hesitate before she rejects this triumph over Carl and "that Miss Meredyth," who, of course, knows that she has usurped Carl's heart. Zai has considered herself bound in honour to Carl; but he himself, by his conduct in the latter days, has given her back the freedom she did not want. There is really nothing to prevent her accepting Lord Delaval except—and that is a great deal—her own wilful rebellious soul, that clings to Carl with a tenacity stronger than herself.

"You will not press me, Lord Delaval! for an answer, will you?" she asks, quietly. "I should like to think a little, to reflect. One can't make up one's mind in a minute you know," she winds up more hastily.

"On condition that you won't keep me too long in suspense. Will you let me know my fate at the State Ball on Friday? That is two whole days."

"Yes," she answers, gravely; then she jumps up from her chair.

"I have promised Percy Rayne, Number 24," she says, examining her ivory tablets, "and I hear it beginning. 24. *Le Premier Baiser.* It is such a delicious air that I never miss it."

He rises and offers his arm in silence.

"It was Rayne who suggested your fancy dress, I suppose? I know he is great at such things," he says, a trifle sullenly.

"Yes; do you like it?"

"No!"

"No! How very rude of you, Lord Delaval! I thought you were the pink of politeness," she replies, laughing.

"I don't like it because I feel as if you belonged to me, and I don't care for you to wear what any other man suggests."

"But I don't belong to you," she blurts out, on the spur of the moment. "Your feelings make a great mistake if they tell you I do."

"They tell me that you will belong to me, however," he answers, in a masterful tone, and Zai feels a thrill pass through her—a thrill of fear almost. It is not the first time she has felt it when this man has had a possessive ring in his voice.

Five minutes afterwards she has thrown off the feeling, and is dancing away as if her heart was as light as her feet; but when the waltz is over, she leans back against the wall, and wishes that she was dead.

"If you have one dance left, Miss Beranger, will you give it to me?" says a voice beside her.

Zai starts, the colour flames into her face, her limbs tremble, and her heart beats so that she places her hand unconsciously on it as if to stay the throbs.

"Yes, I have a dance—this one," she says hurriedly, almost incoherently, and unseen by her people or Lord Delaval, she passes through the swaying crowd on Carlton Conway's arm.

"Come out of the room, Zai, we can't talk here."

Ah! how his voice seems to bring back life and hope and happiness to the love-sick girl. To think! to think! that after all Carl has not thrown her over—that she has been doubting him, doing him injustice all this time.

And as they reach the same corridor in which Lord Delaval has just asked her to be his wife, but passing out of it enter a deserted balcony, the moonbeams fall on her face uplifted to her lover's.

"Once more," Carl murmurs with genuine feeling. "Oh, my love, my own—own love! I have wearied for this!"

And clasping her in his arms, he kisses her—kisses her with the old, old passion—on her sweet lips, that smile and quiver with bliss at his touch.

"It was not true, Carl, what they told me?" she says very low, with her eyes so wistful and one white arm round his neck.

"What did they tell you, Zai?" he asks brokenly. For fickle and light of nature—he cannot look on these sweet wistful eyes—he cannot feel the clinging clasp of this white arm unnerved.

"They told me you were going to marry—Miss Meredyth, Carl."

Her heart throbs so fast he can hear it, but though he knows suspense is a terrible thing, for a few moments Carlton Conway gives no answer.

# CHAPTER VI
# IN THE BALCONY.

"But you!
If you saw with your soul what man am I,
You would praise me at least that my soul all through
Clove to you—loathing the lives that lie.
The souls and lips that are bought and sold,
The smiles of silver and the kisses of gold!"

ZAI looks up hastily at her lover, and her eyes meet his.

It is not only at the touching of the lips that spirits rush together, as many believe. Who has not seen the soul leap up into the eyes, and utter there its immortal language far plainer than mortal speech can interpret it—when pride, or honour, or duty, or interestedness has laid an iron hand across the mouth.

At such a moment we seem to realise with startling force the existence of the divine spark prisoned in its house of clay. The power of spirit over matter, the subtle imagination which, without words, can lay bare

"All thoughts, all passions, all delights,
Whatever stirs this mortal frame."

Before Carl can utter a sentence, he half forgets everything in the sweetness of the grey eyes, in the fairness of his young love's face.

"My darling—my own darling," he whispers, straining her again to his heart, which, to do him justice, he verily believes is devoted to her. "Why have you forgotten me for—Delaval, Zai?"

Zai starts and flushes.

"But I ought not to blame you," he goes on; "after all, class should mate with class, and I am not good enough for you—nor rich enough. I have plenty of shortcomings, I know, Zai, but you must not think worse of me than I deserve."

Her heart flutters like a bird at this, and her eyes glisten through unshed but irrepressible tears.

"Worse of you than you deserve, Carl!" she falters, while her arm clings closer to his neck, and she feels that this man is a king among his kind, and that she may well be forgiven if she worships him. "Why should you imagine that I think any ill of you?"

"Because I merit it after the brutal way I treated you at the Meredyths', and even in the beginning of this evening, my Zai. I doubted you, you see, and when one suffers one is apt to be unreasonable, and wounded vanity is quick to come to the side of wounded love, and after all what is more natural than that you should not love me?" he asks, but clasping her even closer and kissing the bright chesnut hair that gleams up so ruddy under the moonbeams. "What more natural than that you should love—Delaval!"

But in his heart he does not for a moment believe that she or any other woman could pause between any other man and *him.*

"Nothing more natural, I suppose," Zai answers, nestling her hand into his, and feeling her spirits rise and her courage rear its crest aloft as she thinks Carl has only acted thus out of jealousy. "But natural things do not always come to pass, do they? There are exceptions to all rules, you know. I told you before, Carl, that I was the exception to the rule in the Beranger family of being dazzled by Lord Delaval's fascinations. Have you forgotten this?"

"I thought *you* had forgotten it!" Carlton Conway murmurs in his most melodious and reproachful accents.

"Why should you have thought so?" she asks wistfully.

"It would be wiser to ask why I should have thought otherwise," he returns, a little drily. "Your sweet face has bewitched me until I have had no sense left I think, but still I am not *quite* mad. I know my superiors, and am not surprised when fate and fortune compel me to bow to them."

"But Lord Delaval is *not* your superior, Carl!" she cries earnestly, "not in any respect—except that he is a little richer, perhaps."

"I did not mean to imply that he is my superior because he is a swell," he observes rather haughtily, "but the very point of which you speak is the very one that makes his superiority, probably, in *your* eyes."

"In my eyes!" she answers in amazement. "Oh, Carl, I am sorry you should give me credit for such things. I don't think that kind of superiority worth anything—*anything*!" she goes on scornfully. "I don't think that money and position and all that sort of thing makes people really happy!"

"Everyone in Town thinks you mean to make the experiment, anyhow!" he replies.

"But *you* didn't. Surely you didn't, Carl! You know I don't care for Lord Delaval—and that I love *you*!" she whispers, *les larmes au voix*.

He looks down at her sweet downcast face. It is a face bathed in blushes. For Zai always blushes when she tells him all that is in her heart. But she need say nothing. He has only to look at her face, which tells its story of love with exceeding clearness and sweetness to his vain, incense-loving eyes.

"Zai! do you really love me so very much?"

He asks the question from sheer selfishness and a desire for incense to his overweening vanity. He knows he has sought this opportunity to tell her something which will break her heart. But no—hearts are tough things, and do not break easily. But something which will surely wreck her implicit child-like faith in the fidelity and sincerity of all men. Never after to-night will Zai Beranger perhaps feel that loving words and honest words are twins. Rather she will shrink from them, knowing that they may be uttered only to betray.

Now she believes in Carlton Conway with her whole soul. And when he asks:

"Zai! do you really love me so very much?"

She lets both white arms form a circle for his neck, and woos him to touch her red lips.

For one moment she forgets her maidenly reserve, and only remembers that in her own eyes she is his wife—in heart, if not in name.

"Oh Carl! Carl! let us marry at once—dear! and then no one can come between us two!"

"We cannot!" he says hastily.

Zai starts as if she were shot, and covers her face with her two little hands, while a burning blush surges over it.

It comes to her suddenly, the terrible, terrible shame, of *her* having asked—of *his* rejection—and then the colour leaves her cheek.

She leans against the balustrade, with the moonlight falling on a face white as undriven snow. Her eyes have a dumb misery in their depths, and her mouth quivers like a child's.

"Oh Zai! forgive me if I hurt you by saying we cannot marry!" he whispers brokenly, for her white face and trembling lips move him strangely, worldling as he is. "You know very well how I am placed! I have nothing but my salary, and that is dependent on health; and if I don't marry some girl with money, I don't know what will become of me, Zai!"

A deep silence ensues for a minute or two. Up above the glorious moon sails serenely along, and a few feathery clouds float athwart the great sapphire plain of sky. From within, the sound of music is carried out on the fragrant night, but human eyes and human voices are nowhere near.

These two are alone, entirely alone, on this isolated balcony, and they have for many months played at making love.

Listen then in what passionate words Belgravians and worldlings say farewell, if farewell must be said by them.

We all know that Romeo and Juliet would not have said it, but they were foolish inconsequent young people, who fortunately did not live to test the agreeabilities of a narrow income.

"Then I suppose you are going to marry Miss Meredyth?" Zai asks in a low voice, that has a hardness in it which no one has heard before.

"Zai! can you blame me? Can you think it possible for me to act otherwise?"

"No! I don't blame you!" and again bitterness mars the sweet voice.

"Of course you cannot blame me!" he answers, "for you know *you* are forbidden fruit, Zai. You have been reared in certain social conditions, which of course it would be sheer wickedness on my part to ask you to resign!"

This is a very different sentiment to what he has expressed before; and even she, much as she loves him, feels indignant.

There is a sudden flash in her grey eyes as she lifts them to his.

"You know that you ought not to say this, Carl! It is not my interests you are thinking of, but you have made up your mind not to marry anyone who has no money!"

"Granted!" he replies quietly, though a crimson flush dyes his face, and he bites his lip hard. "But though you seem to reproach me, you know why it is so! You know that people in *your* world cannot subsist on sentiment, or on a few paltry hundreds a year. I am, I avow, one of those miserable devils to whom the bitter irony of fate has given the tastes and habits of a gentleman, without the means of supporting them. You are the corresponding woman. Common sense—the commonest sense—will tell you whether or not it would be sheer madness for us two to marry, although we love each other so passionately, Zai!"

Zai does not answer. There cannot be the least doubt, she knows, but that common sense *does* tell her that marriage with her would not suit Carl Conway; but it is none the less true that common sense is not what she cares to listen to now. In the most vapid soul that sojourn in Belgravia ever starved, there is still some small lodging left for that divine folly that men call "Love."

And Zai, born and bred in Belgravia, is as desperately and honestly in love with this man, who has played fast and loose with her, as a milk-maid could be.

She longs—how she longs—for just one crumb of comfort, just one little word of sweetness from his lips.

Only a quarter of an hour ago he held her to him and kissed her with apparently the old, old passion in his soul, and now he stands a little apart, calm and cold as a statue.

Conway is a wonderfully handsome man, and Zai worships his beauty. The more she looks at him the more she craves for a gleam of love in his brown eyes—the stronger grows her desire to listen to love from his well-cut lips; but she listens in vain.

"Yes, I know all that," she says very wearily, with a dreadfully heart-sick feeling of disappointment, "it was hardly worth while you telling me. I have heard papa and mamma, and Gabrielle, and all the others talk of 'common sense,' but one grows tired sometimes of hearing the same thing."

The tone of her voice tells more than her words; there is a betraying quiver in it that makes him turn quickly and look at her.

The eyes that meet his own have great glittering tears in them. Never in her life has Zai looked more lovely or more lovable than at this moment, and Carl recognises fully all that he is sacrificing for money.

"Forgive me for having repeated anything then that wearies you," he says softly, clasping her cold white hand in his own, and Zai lets him. Even now—even now! in spite of his falsity—his avariciousness—the touch of his hand thrills her through and through, and her white lissom fingers linger in his grasp. "Zai, my darling! you *must* feel that it is as hard—much more hard indeed—for me to utter than for you to hear. Good Heavens! do you imagine I am thinking of myself? (For a moment, perhaps, he really fancies he is not.) It is of you, my dearest, that I think. How can I be so cruel—so selfish as to ask you to give up for me everything that you have been taught all your life to consider worth possessing? But if you really wish to do so, Zai, I can only say that you will make me very happy. And, darling, you know I shall strive very earnestly to keep you from regretting it!"

Brave words these are and bravely spoken, with not a single falter in the tone—not a sign of what they cost, but a swift pallor sweeping across his face.

Let us do this worldling credit—let us confess that it is very well done for a man to whom nothing could be more ruinous than to be taken at his word.

But frankly, Carlton Conway has not reckoned without his host. It is a curious rather than an absurd sense of honour that forces him to risk this declaration; but he knows the girl beside him too well not to be *almost* certain of her reply.

The event justifies the expectation. Zai loves him to distraction, and the loss of him will create a void in her life which she believes no one on this earth will fill up—not if she lives to be as old as Mount Horeb.

Carl's handsome captivating face tempts her—the most genuine love that a woman can feel tempts her to keep him at any cost.

But it is only for a moment she wavers.

She knows that Mammon and Cupid have run a race in Carl's heart and that the former has beat by several lengths.

Young, ignorant of guile, and innocent, a sort of instinct teaches her this.

"It is impossible!" she falters, with the sharp thrill in her soul echoing in her voice. "You are perfectly right, Carl, in all you have said, and I—I know it as well as you do. I have been reared under certain conditions and for certain ends, and perhaps I could not put them entirely aside. I am fit for nothing but Society, and Society would not recognise me if I was poor and struggling, so we should simply mar each other's lives and render each other miserable. And, Carl," she tries to speak calmly but the effort is terrible, "I could not bear poverty and neither can you, though—— " She breaks down completely, large tears chase one another down her cheeks, but she dashes them away, wroth at herself for her weakness and want of pride. "Therefore we must not think of marrying, of course!"

Another dead pause. Madam Diana sails along more brilliantly than before, this time with an enormous court of glittering stars around her. The cool night air passes quietly by, lifting up the chesnut tendrils of hair that stray on to Zai's brow and fanning her poor hot temples. The time is flying by, and someone will be coming this way, but nevertheless Carlton Conway cannot end this interview without a few more words.

"And you will of course let Lady Beranger persuade you into marrying Delaval?" he asks, jealously—angrily.

Like the dog in the manger, he does not want the girl himself but he grudges her to another man.

Jealousy is a passion that is often wonderfully independent of the passion of true love.

Carl is very loth indeed that Lord Delaval, whom he has always hated, shall have this lovely piece of nature's handiwork for his.

"I don't know," Zai murmurs wearily. Then she calls up all the high spirit she has in her and says quietly—"After all, the matter might be worse—for Lord Delaval everyone says is charming, you know."

"But you care nothing for him, Zai! You care for *me*!" he exclaims passionately, with almost a mind to claim her sooner than she should pass out of his life in this manner.

"I know—and yet—— "

"And yet you *may* become Countess of Delaval?"

"I may."

Upon this Carl releases her hand pettishly and subsides into silence. He is not of a nature to ponder deeply on social or any other kind of evils, but just now the sordidness of this strikes him very forcibly, and he wonders how such girls as the Berangers hold themselves even a degree better than the Circassian and Eastern females who sell themselves for filthy lucre.

"Zai, tell me the honest truth. Do you care for Delaval the least bit in the world?" he asks earnestly, longing for her to deny the existence of any liking for his rival, to protest the enormous height and depth and width of her love for himself.

"Not yet—but," Zai adds slowly and meditatively, "if I marry him I shall do my best to care for him, and even if I didn't—what of it? Do people in our world deem it necessary to care for the man or the woman whom they marry?"

And Carl Conway cannot honestly affirm that they do.

# CHAPTER VII
# THE STATE BALL.

"I have hidden my soul out of sight and said
Let none take pity upon thee. None
Comfort thy crying—for lo! thou art dead.
Lie still now, safe out of the sight of the sun;
Have I not built thee a grave, and wrought
Thy grave-clothes on thee of grievous thought?"

THE June sun is full of pranks to-day. There it is, scorching up the leaves in the square, broiling the toilers on the white pavements, shining down on everything with a lurid glare that makes one wink and blink, and generally uncomfortable, and now it is peering into the windows of Baby's schoolroom, showing up the short-comings of the faded carpet, the ink stains on the old table, and streaming full on to a corner where, before her easel, Zai stands, palette and brush in hand, but idle.

"Oh, it is hot! hot!" she cries impatiently, throwing down her painting apparatus and pushing her hair back from her forehead.

"Here's something to cool you!" Gabrielle says, throwing across the *Morning Post*, and then she has the good feeling to pick up a book and pretend to be buried in its contents, while Zai reads what she considers her death warrant.

"A marriage has been arranged and will shortly take place between Miss Meredyth, daughter of John Meredyth, Esq., of Eaton Place, and Carlton Conway, Esq."

Three times Zai reads the announcement over—mechanically spelling each word—then she drops the paper on the floor, and going up to the open window, looks out.

She does not find the sun hot now, although it is dancing on her chesnut hair, and turning each tress to fire. Her heart lies so dreadfully cold within her breast that it seems to ice her whole frame, and though her eyes face the strong yellow beams, they do not shrink from them.

Since she read the words in to-day's *Post*, she seems to be blind and deaf to everything, save the fact that Miss Meredyth has won from her that which she valued most in life.

"Well, Zai?"

Zai has been standing at the window perfectly motionless for half an hour, her slight figure almost rigid, her head a little thrown back, her face white as marble and almost as impassive, her two little hands clasped behind her as in a vice, and Gabrielle thinks it high time to recall her to a sense of everyday life with all its ills.

"Well, Gabrielle!"

The girl turns and faces her step-sister; her eyes look as if she were stunned, but her lips smile.

Gabrielle stares at her for a moment, then she bends over her volume again.

"There, child, don't act with only me for an audience!" she says quietly, "You have had enough of acting and actors, goodness knows. What a brute the man has been!"

"Why?" Zai asks defiantly.

"Why?—because he pretended to love you, and he *knew* you loved him, and yet he has quietly bowled you over for that doll of a thing."

"He cannot help himself, Gabrielle!"

"Why cannot he help himself, pray?"

"Because Carl is so poor. Oh, Gabrielle! Gabrielle!" and, the tension passed, Zai throws herself down on Baby's favourite hearth-rug and sobs as if her heart would burst. "What an awful, *awful* thing money is!"

"The *want* of it, you mean! But that man Conway knew he was poor always. Why did he ever spoon you as he has done?"

"He loved me so—he could not help it!" Zai says tenderly, "And we love each other dreadfully—*dreadfully*—still, but he thinks I should suffer so if I did not have the luxury I have been accustomed to all my life!"

"And he does not think about himself, poor dear unselfish fellow!" Gabrielle says with a little sneer. "Zai, take my advice, and don't waste another thought on him. He is going to marry Miss Meredyth for her money, let him, and don't let Miss Meredyth have the pleasure of seeing that you envy her her husband!"

"I must try and forget Carl," Zai murmurs feebly. "It would be a sin to love him when he is married, but I don't know how to begin. He seems to run in my head and my heart so!"

"Let some other *genus homo* turn him out of them. There's heaps of eligibles about. Lord Walsingham, for instance, he is young, good-looking and tolerably well off."

"Why he squints, Gabrielle! and has red hair!" Zai protests mildly.

"Never mind. What does it matter whether one's husband has red hair and a squint? All one wants is a nice house, and fine carriages and horses, plenty of diamonds etc. Is there no other man you know who could make you forget that actor fellow?"

"No one!"

Zai blushes crimson. There is meaning lurking in Gabrielle's manner and eyes, although her words are simple enough, and she remembers that this step-sister of hers has resolved to win Lord Delaval for herself.

Let her, Zai thinks; she has never felt so much distaste to accepting Lord Delaval's offer as she does at this moment, when her heart is so sore and her spirit so humiliated.

"I won't cry any more!" she exclaims, feigning to be indifferent, but in reality anxious to change the subject. "I must look well before the Royalties to-night, you know! The Prince was very nice to me at Caryllon House, and said I was the belle of the room! What are you going to wear, Gabrielle?"

"Black lace—and you, I suppose, are going to wear sackcloth and ashes!"

"No I am not!" Zai answers lightly. "Mamma coaxed Swaebe out of another six months' credit, and so Trixy and Baby and I have loves of pale blue faille and white illusion, and water lilies trailing all over us. I want to look beautiful to-night for a reason.

"What reason?" Gabrielle asks, suspiciously.

"Only because—— But no; it's a secret for the present." And Zai, running out hastily, rushes up to her bedroom, and, double locking her door, cries to her heart's content.

They are about the last tears dedicated to the memory of Carlton Conway; but, by-and-by, she bathes her eyes in cold water and smoothes her hair, and putting on her hat, goes out into the Square. But the Square is associated in her mind indelibly with that evening when she stole out from Lady Beranger's ball to meet her faithless lover, and rising hastily from the bench, she walks home again.

"Go and lie down, Zai, and rest yourself; you look like a ghost!" Lady Beranger says harshly, meeting her on the stairs. "Or better still, put on your white chip hat with the pink roses, and come with me to the Park. The air will beautify you, perhaps."

And Zai—who has learned by this time that Lady Beranger's suggestions are really fiats—goes up and adorns herself, and is quite bewitching in the chip and roses by the time the Victoria is at the door.

Lady Beranger leans back, a trifle pale, and with the *soupçon* of a frown on her brow, and the carriage is just at Hyde Park Gate before she volunteers a remark.

"You have seen the *Post* to-day?" she says, carelessly.

"Yes, Mamma, and I am so glad to see Mr. Conway is going to be married; Crystal Meredyth is very nice, and awfully rich, you know."

Lady Beranger turns round slowly and fixes her keen searching eyes on her daughter.

But Zai has not been born and bred in Belgravia for nothing.

Not a lash quivers—not a change of colour comes—under the scrutiny.

"I always said Carlton Conway was a cad!" her ladyship observes coldly; "and I am very glad you have found it out too."

"But I haven't, Mamma, not the least in the world. I think quite as well of Mr. Conway as ever."

Zai's self-possession amazes and almost annoys Lady Beranger. She is positively out-Heroding Herod! But she only says, in a cold, hard voice:

"Think as well of him as you like, Zai, so long as you keep it to yourself. His sort of people are all very nice in their proper places, but I have never advocated their being in Society. There *is* the individual in question!"

Zai looks eagerly round, and her cheeks glow crimson and then wax pale, and she bites her lips to stay their trembling, as the Meredyths' high Barouche with stepping roans dashes by, having for its freight only Miss Meredyth and her *fiancé*! (Mrs. Meredyth, not so scrupulous as Lady Beranger about the *bienséances*, thinks there is no harm in an engaged couple being seen alone in the Park.)

Miss Meredyth, dressed in rose colour, with a sailor's hat perched coquettishly on her fair hair, looks uncommonly pretty, and so Carlton Conway seems to think, for he is so engrossed in regarding her that the Berangers' Victoria is passed unnoticed.

"I thought it was the Meredyth girl's money the man was after, but he seems to be *énormément épris*," Lady Beranger remarks indifferently, hoping the shaft will fly straight home and cure all remaining nonsense in her daughter's head, or heart, or wherever it may be.

Zai answers nothing. With a sharp pang of misery and jealousy, she, too, has noticed how devoted Carl seems. *Après cela le Déluge.*

She is thankful when her mother orders "Home." She is sick of bowing and smiling when she would like to lie down and die; but nevertheless she trips airily down to the dining-room, eats more dinner than is her habit, and after this goes into the conservatory and plucks a couple of the reddest roses she can find.

"Fanchette, make me awfully pretty to-night!" she coaxes, and the *femme de chambre* is nothing loth. Zai has every "possibility," as she calls it, of being *belle comme un ange*, and more than satisfies her exquisite Parisian taste when her toilette is complete.

"She wants but two little wings to make her a veritable angel," Fanchette says to the English maid who assists her in her duties. "Mees Zai is the flower of the house!"

"Flower of the *flock*, you mean," Jane corrects.

"No, I do not," Fanchette replies, offended. "I have *never* heard of flowers in a flock. I have heard of a flock of goose—and *you* are one of them."

Meanwhile, Zai stands before her mirror. Her eyes are so sad—so sad, that they look too large for her small white face.

"Oh, Carl! Carl!" she says, half aloud, "you have forgotten me quite! And I love you—love you so much that my heart is broken, Carl!"

"Zai, the carriage is ready," cries Baby, drumming her knuckles on the closed door.

Zai starts guiltily. What right has she to be murmuring love words to a man who will soon be another woman's husband!

She clasps a pearl necklace round her throat, fastens a pearl star into her bonnie brown hair, then pauses one moment.

It is the first time in her life that she has ever had recourse to the foreign aid of ornament, and it seems quite an awful thing to her. But no one must guess at her feelings from her wan face to-night. She had not been proud with Carl because she loved him so, but she must be proud with the world, and not wear her poor desolate heart on her sleeve for daws to peck at.

She takes the two roses she plucked, pulls off their petals mercilessly, then rubs them on her cheeks, and flinging on her cloak she runs downstairs.

Lady Beranger is putting the finishing touches to her elaborate dress of primrose satin and *point de Flandre*, in which she looks like an empress, and only the three girls are assembled in the hall when Zai appears.

"How do I look?" she asks, throwing off her wrap. "Fanchette says I look *belle comme un ange*, and I want to be especially beautiful to-night!"

"What for?" three voices ask at once. "It's only a State Ball, on the pattern of all the others we have been to. The Queen won't be there to make anything different. So what on earth does it signify how you look?"

"I'll tell you!" Zai says slowly and deliberately and unflinchingly. The rose petals hide the pallor on her cheeks, and the smile on her lips does away with the sadness in her eyes. "But, girls, you must keep it a secret from the Governor and Mamma. I want to look my very best to-night, because I intend to make my bow before the Princess as a future *Peeress*!"

Lady Beranger enters at this moment.

The State Ball is worth seeing after all, though the Beranger girls had said that it was exactly on the same pattern as its predecessors, and that Her Gracious Majesty was not going to shed the light of her august presence to make it any different.

Seldom within four walls has more beauty been gathered than to-night. Of course everyone admires the Princess most, but of feminine loveliness there is every possible variety to suit every possible taste.

There is also a good deal of the feminine element which is not lovely. But, as if to atone for Dame Nature's shortcomings, it is generally expensively dressed.

Zai soon has cause to forget or despise Fanchette's soothing doctrine of the fitness of things, and to feel that her pale blue *faille* and white illusion, garnished with water lilies, are chiefly remarkable for their fresh simplicity, as she views the superb silks and satins and laces that do honour to Royalty.

She dances away with half-a-dozen of the Household Brigade, with the Duke of Shortland, Lord Walsingham, and several Belgravian *habitués*, and then she walks through the room with Percy Rayne.

He is quite as good as a catalogue in a ball-room. Ever since he was a small boy Fate has hung him about the Court of St. James'. He has the names of the upper current, and all the social celebrities, on the tips of his well-shaped nails, and faces he never forgets. Added to these, he has all the fashionable gossip on his tongue, for in the interludes of "business" at the F.O., as well as at the other "O's," they enjoy a dish of scandal as much as the softer sex do.

He points out the Beauties now to Zai, who, in spite of her heart-broken condition, regards them with admiring interest.

"There!" he says, "is an American, Mrs. Washington Ulysses Trotter, called the Destroying Angel, because she kills everyone dead, from Princes downward, by a glance of her beautiful eyes; but, unfortunately for her, her triumphal car will be probably stopped in its career. The Yankees are going out of fashion, you know. Royalty has decreed it. For Royalty, like common flesh, is liable to get bothered with being run after and accosted as if it were Jack or Tom or Harry. But Mrs. Washington Ulysses Trotter does not mind much. She knows her little outing at Buckingham Palace is quite enough to get her the *entrée* into all the Fifth Avenue houses. She will talk about the Prince—

"Oh my, isn't he elegant, and so chatty! I felt just like talking to Cyrus Hercules Hopkins—that's my cousin down Chicago way, you know. And the Princess! well, certainly, *she* isn't proud! It was just like being at home in our English basement brown stone house, Maddison Avenue—at Buckingham Palace!"

Zai laughs, and he rattles on.

"That's one of our big financier's daughters. Ugly, isn't she? I hate the type. The *parure* of brilliants isn't bad, and those yards of lace—*point D' Alençon*, isn't it—that trail about her are worth more than my year's salary. But they are so devilish stingy in the Offices. We work like slaves, and get neither tin nor *kudös*. And you would not believe it, Zai, but the Foreign Secretary hasn't more responsibility on his back than I have on mine! See! there's the famous wife of one of the Ministers—Count Schoen. She has been a celebrated beauty in her day, and cannot forget it. And they say she enamels and bakes her face in an oven. What do you think a cousin of mine—an *ingénue* from the country—did, at the Caledonian Ball? She went up to the end of the room, and after intently examining Count and Countess Schoen, said aloud,

" 'How funny that they have Madame Tussaud's figures here.'

"Imagine the horror of her partner!"

Zai laughs again. But this time the laugh is forced, and she catches her breath hard.

Through the swaying crowd she espies Gabrielle among the bevy of beauties.

Gabrielle holds her own to-night. Her black lace dress becomes her white creamy skin admirably. Scarlet japonicas burn and gleam in her coal-black hair and on her bosom. On her cheeks, the bright pink flush lends increased lustre to her large dark eyes. As she sweeps along she has that supreme unconsciousness of manner which is never seen save in a woman who feels she is well dressed and able to defy the criticism of her own sex.

Gabrielle does not see Zai or Percy Rayne looking at her, for her eyes are mostly cast down on the fan she carries, neither does Lord Delaval, on whose arm she leans, observe them, for he is bending and speaking very low under the sweep of his long fair moustache, while his glance rests on the undeniably very handsome face near his shoulder.

"Don't they make a good looking couple?" asks Rayne. "What a pity they don't arrange to walk through life together—they look so well doing it through a ball-room."

"They are both handsome," Zai answers indifferently, but she is, spite of her, a little piqued.

This man—to whom her answer has to be given to-night—has not even deemed it worth his while to ask for it, though the evening is wearing on. His neglect hurts her more, sore and suffering so lately from Carlton Conway's behaviour, and poor little Zai feels that she would like to hide her diminished head for ever.

"I am very tired," she says to her partner; "Do you think I could get a seat somewhere?"

"Yes; but come out of this crowd. It's awfully hot, and you look like the whitest lily, Zai—we'll find a seat somewhere."

So they go out, and he finds a chair for her in a vestibule, where a little cool air revives her.

"I *must* go. I have to dance this with Lady Vernon. Do you mind sitting here quietly till I come back?" he asks kindly, seeing how weary and wan she looks.

"I should like to stay quiet here very much," Zai answers gratefully; "and don't hurry back for me."

She half closes her eyes, and fans herself slowly, and feels desolate—so desolate.

Her womanly triumph over Miss Meredyth has evidently fallen to the ground; Lord Delaval has either changed his mind, or else he was only laughing at her at Caryllon House—and as she thinks thus, Zai shivers with mortification and shame, and leaning her head against the wall, grows lost to external things.

She does not know how long she has sat here, and she does not care—all she yearns for is the solitude of her own room; but the ball is not half over, and hours—dreary hours—lie before her.

"Zai! is it to be—Yes?"

She starts up, flushing red as a rose—her heart beating wildly, her eyes with a dumb wonder in them.

She is but a bit of a girl, she has been cruelly jilted by the man she loves, and she craves for a little incense to her *amour propre*, even though it be dearly bought.

"It is—yes," she almost whispers; then in a sort of mist she sees Lord Delaval's face light up, and the colour creeps warmly over his blond skin.

"Thank you, my darling!" he says very low, bending over her, and she feels his lips touch her bare shoulder. Then she puts her hand on his arm, and without another word they walk back into the ball-room, and up to Lady Beranger.

"Let me present to you the future Lady Delaval!" he says quietly, and Zai slips her ice-cold fingers into her mother's clasp, and for the first time her mother looks at her with positive affection in her glance.

"Is it true, Zai!" she asks, eagerly.

"Quite true, Mamma," Zai answers without a falter.

A little later the news has been told to the Royalties, and with kindly smiles and words they give their congratulations on her future happiness.

But though the Royalties know of the match in prospective, Zai pleads that it may be kept a secret from her sisters for the present. It may be that the death and burial of her first love is too recent to permit of matrimonial rejoicings just now, or it may be that she wants to realise what has come to pass, and to resign herself to the future before the others touch upon the subject, and probe not too quietly the still open wound made by Carlton Conway. Lord and Lady Beranger are too well pleased that matters have turned out so satisfactorily to refuse her request.

And, as for Lord Delaval himself, perhaps he feels a little uncomfortable at appearing on the scene as a devoted lover before Gabrielle—Gabrielle, who has told him, in the passionate words that rush unchecked to her scarlet lips, that the day of his marriage to any other woman will be the day of her death.

She is not one to kill herself; she is not romantic enough for folly of that kind; what she means is probably a social and moral death; but Lord Delaval—with the innate vanity of his sex—believes that Gabrielle's handsome face and superb figure will be found floating on the turbid bosom of old Father Thames, and he shrinks more from the scandal of the thing than from the remorse likely to rise up in his breast. Zai's desire, then, that the engagement shall be kept quiet for a while, meets with his approval. After all, he can find chances to gather honey (if not all the day) from his betrothed's sweet lips—and stolen sweets have always been nicer to his thinking than any others.

When they say good-night, he contents himself by squeezing five very cold fingers, and slipping a magnificent brilliant on to the third one, which pledge of her bondage Zai does not even glance at before she drops it into her pocket.

"Did you like the ball, Zai?" Trixy asks, as they brush their hair before going to bed.

"I hated it," Zai answers, giving her chesnut tresses an impatient pull. "I wish I had never gone to it!"

# CHAPTER VIII
# "SIMPLE FAITH THAN NORMAN BLOOD."

"You'll look at least on love's remains;
A grave's one violet!
Your look? that soothes a thousand pains.
What's Death? You'll love me yet!"

"JUST be careful who mounts that chesnut to-day, Hargreaves," Challen, the riding-master, says, pausing on his way at the door of the stable, and passing a keen glance over the horse in question. The chesnut is a big, good-looking hack, with a sleek satin coat, and just what would take a woman's fancy, but there is a look about his eye that Challen does not like. "Put Miss Edwards on him, she has pluck enough to ride to the devil, but mind none of the new pupils go near him."

Hargreaves assents, but he does not look content.

"She wants to ride the chesnut," he says to himself. "She's set her mind on it, and I hate to disappoint her! Bless her heart! Why, what's the matter with you?" he continues aloud, going up to the chesnut, and passing his hand over the long, lean head. "I like you, because she likes you! You'd never think of hurting her, I'll be bound, no more than anyone would, I know! My pretty one! I'd kill myself if any harm came to you—that I would!"

And Gladstone Beaconsfield Hargreaves, quasi village veterinary, but now assistant-master of the Belgravian riding-school, pulls out a tiny locket from his breast and kisses it a dozen times, then holds it up to the light reverentially as if it was the holiest thing to him on earth.

"Just like a bit of gold it is, for all the world! The same colour that angels' hair is. Oh! my pretty one; my sweet one! There's never a night I don't go down on my knees and thank God that you don't scorn me!"

It is the morning after the State Ball, and while the other Beranger girls take an extra hour or two of slumber, Baby, fresh as a lark, dons her dark-blue habit that fits her lovely little figure like wax—and is off for a riding lesson.

The weather is true summer, and the little lazy breeze that floats across the Serpentine is a boon to man and beast. Right away in the upper portion of Kensington Gardens, the trees throw down some grateful shade, and Challen's riding-school wend their way down the broad walk at a snail's pace, for the heat is awful.

Up above there is not even a cloudlet to temper the sun's rays; the sky is as clear and as blue as Baby's own eyes, and everything around looks as bright as her smiles.

There are not as many aspirants to equestrian honours as usual to-day. The season is on the wane, and the Ball and Reception givers pile on the agony fast and strong, so that the young *débutantes*, fagged and worn out by nocturnal exertions, find the arms of Morpheus more to their liking than the caresses of Boreas.

Miss Juliana Edwards, a strong-minded, steel-nerved brunette, and Challen's show pupil, is here, well to the front of the small cavalcade, but she does not ride the chesnut.

Her dare-devil propensities find but small play, for her mount is a dapple-grey gelding, who looks as if neither whip nor spur will rouse him out of riding-school jog-trot.

There are only eight riders in all, and the first lot go in threes, while some little distance in the rear Hargreaves keeps close to the chesnut, on whose back is Baby.

"You'll kindly look to the other ladies, Miss Edwards, won't you?" he had said on starting, with a deprecatory smile. "I think I had better keep an eye to Miss Mirabelle Beranger's horse. She doesn't ride like *you* do, you know!"

And Miss Juliana Edwards, to whom a compliment on her horsemanship is dearer than anything, smiles in return at the handsome assistant, and agrees to keep a sharp look-out.

The chesnut goes steadily enough—so steadily in fact, that Baby, who is an awful little coward, forgets all about him, and gives her whole attention to her teacher, who, in the neatest of grey tweed suits, and with an unimpeachable wide-a-wake perched jauntily on his curly head, looks quite *the gentleman*.

"I *wish* you had been at the State Ball last night!" she says, with a beaming smile, that almost takes the young fellow's breath away.

"*I!* fancy me at a State Ball, Miss Mirabelle!"

"Why not? I am sure there was no one so good-looking as you there!" she cries, looking admiringly at the trim, slight figure, and the straight features and undeniably winsome eyes of her companion. "I *wish* you would not call me *Miss* Mirabelle!" she adds with a little pout of her charming red lips.

He reddens visibly as he hearkens.

"I *dare* not call you anything else, Miss Mirabelle!" he almost whispers, his heart throbbing violently under his tweed waistcoat.

"There it is again! *Miss* Mirabelle! why can't you say 'Mirabelle,' when—when—we are *quite* alone?" she asks impatiently, throwing a covert glance towards the other riders to see if they are out of earshot.

"Oh! *I couldn't!*" he murmurs very low—shy of speech—but his large hazel eyes are eloquent enough. "I would as soon think of calling the angels by their names!" he goes on nervously.

"I have heard of Michael as the name of an archangel, but I don't think the female angels *have* any names," Baby says irreverently. "Do you think *me* an angel? because I'm not, not the very least bit in the world. The governor calls me a little devil, and I know my sisters don't think me an angel!" she laughs.

"You are an angel to *me*, anyhow!"

A little pause, while she looks straight into his eyes, with the prettiest, faintest pink colour creeping over her cheeks.

"I say, Hargreaves, how long are we going on like this?" she asks abruptly.

He gazes at her amazed, and Baby laughs again, a little, low, musical laugh that entrances him.

"I mean that—that—as we care for one another, why should we pretend not to?" she asks in a hushed voice, putting her hand on her pommel, for the chesnut pricks up his ears and frightens her. Hargreaves' hand is on hers in a second. He is really rather nervous about the horse after Challen's warning, and besides, it is Heaven to him to feel the soft velvety skin of the dainty little hand that gleams up like a morsel of alabaster statuary under the sunlight.

"Miss Mirabelle, for God's sake don't go and make me forget what I am. I try night and day to remember the distance between us, and though I could go down on my knees and worship you all my life—though I could die for you willingly—*willingly*, I know I dare not *live* for you! I love you—*there!* Only God knows *how* I love you, but it isn't a love like a fellow gives to his sweetheart! It's a love like a faithful dog, that would lick your pretty hand and be content; that would watch over you so that no harm came near you; that would just lie down and die by the side of *your* grave."

Baby listens with an involuntary tear twinkling in her eye. She is only seventeen, but she has been too long in a Belgravian world not to know that this young fellow loves her with a beautiful, unselfish, honest love—the like of which no Belgravian fine gentleman would feel. This primitive, self-abnegatory sort of courtship is so novel that it has a glamour for her, and Baby is—undoubtedly—a little fast.

"I would rather *live* and find out how much you *do* love me, Hargreaves," she answers, with a tender smile; "do you think you love me to—to—the extent—of—marrying me?"

"Miss Mirabelle!" he gasps.

The veins swell on his forehead, his eyes fix on her with a bewildered look, and his breath comes quick and fast. Then he droops his head, and a forlorn expression sweeps over his white face.

"Don't laugh at me, for my dead mother's sake," he whispers in a hoarse tone.

"I am *not* laughing," she says slowly, "not laughing one little bit, Hargreaves. Would you think it very fast of me if I said something—something quite out of the way, you know?"

"I could not think ill of you, no matter what came," he replies earnestly.

"Well then, here goes! I am ready to be Mrs. Hargreaves as soon as you like."

He stares at her like a man in a dream, and as he lifts his eyes to her lovely little face, Baby's snowy lids droop over her cerulean orbs, while her mouth twitches with something between a quiver and a smile.

He is not a gentleman born and bred, but he has a heart that can love. Blue blood may not flow in his veins, but honest, devoted, even chivalric feelings live in his breast, and he *knows* that this girl—in spite of the words she has just spoken—is a thing he dare not grasp.

No, if her love and her presence are Heaven, the loss of her undying misery and regret, he does not dream of hesitating between them for her dear sake.

She has offered herself to him—the sweetest, most precious gift he could have on earth—but sooner than take her, sooner than drag his dainty high-born darling down to his own level, he would shoot himself.

"No, no, Miss Mirabelle! I should be a rascal, a cur, if I thought you were in earnest. I have no right to love you; but love is a thing that comes alike to all, and I may feel it so long as I don't let it harm you, Miss Mirabelle. God bless you for liking me, for speaking to me kindly; but I ask no more than that—only—*only*—may I just kiss your hand—*once*—Miss Mirabelle."

He raises a white, stricken face as he speaks. He has made up his mind to throw up his situation this very night and to go away—to America—Australia—anywhere so that she may never see him again, and regret perhaps that she has spoken to him thus. He will pass right away out of her life, but he wants one kiss of her little white hand to take away with him; that kiss and the locket that holds a bit of her shining hair—his two priceless treasures.

Baby's eyes are full of tears now. The young fellow's voice has such a ring of pathos in it—a ring she has never heard in the voices of Belgravia—but she says nothing, only pulls off the gauntlet from her right hand and holds it towards him.

"*Good-bye*," he whispers so incoherently that she doesn't catch the word, and stooping, Hargreaves fastens his trembling lips on the soft white flesh, *when*——

The Chesnut has started forward, and, off her guard and terrified out of her senses, his hapless rider loses all presence of mind and clings on as the horse careers madly along.

The rest of the school have turned to the right and disappeared from view. Hargreaves, horror-struck, almost stunned, does not follow for a moment, and only the Chesnut with its helpless burden dashes on and on. Turning sharply to the left he gallops furiously—so furiously that all obstacles give way before him. On and on, on and on! till the gardens are left long behind, and the road by the Park is reached, while the poor pale little rider clings desperately on with all her might and main for dear life.

Suddenly the horse swerves to the right down a narrow street, and losing her hold, the girl falls off.

Pray God that the horror of her fate is over! *but no!*

The tiny foot is entangled in the stirrup, and for nearly thirty yards the brute drags her along, when all at once he stops dead short, frightened and quivering, and the jerk snaps the stirrup leather in two.

*But it is a little too late!*

They pick her up, a little white dainty thing. Her hat has fallen off, and her long hair—angels' hair, as Hargreaves has called it—streams down in such long rich shining waves that it seems to envelop the small slender figure in an armour of burnished gold.

She is not dead—her blue eyes, blue as the sunny sky—are quite wide open, and some one, a slight young fellow, who has just ridden breathlessly up, falls down prone on his shaking knees and looks into them with the poor piteous look of a faithful hound.

"Miss Mirabelle, Miss Mirabelle!" he calls in wild despairing tones.

But she cannot rebuke him now for his formal address, poor little soul!

Presently her eyelids droop, and the long curling lashes rest close against cheeks that are almost ashy now.

They lift her up gently and carry her—"Home," the home she had left only two hours before gay and blithesome as a bird and so full of life, and when it is reached they take her straight into the library, the door of which is ajar, and laying her down on the couch, they leave her, all but one, and he does not enter the room that contains her, but stands trembling near the threshold.

Another moment and the awful thing that has happened is known to all in the house, and Hargreaves shrinks away still further as father, mother, sisters of the girl he loves pass him with scared faces and stricken hearts to find Baby—*so!*

Not a word is spoken. At such a moment what word can be said? Even Lady Beranger bows her proud head beneath the fiat of Heaven, while Lord Beranger sobs aloud over this little one—this brightest, merriest one of all the flock.

After a moment, revived by a stimulant, Baby opens her pretty blue eyes.

"Don't cry, governor!" she says in a voice so faint—*so faint!*—that it seems to come already from that distant shore. "It serves me right! I was going to leave you—I was—— "

She stops, struggling for breath.

"Let me just see her, my lady! Oh, for God's sake let me just go near her! I won't dare to touch her—I won't even dare to say *good-bye*!" a voice whispers so hoarsely, so brokenly, that my lady starts and turns round, but does not understand.

But Baby has heard, through the faint mists that are rising up around her; the voice of the man who loves her finds an echo in her heart.

"*Let* him come near, governor," she says slowly, with an effort. "He isn't a gentleman, but I loved him and asked him to marry me, but he wouldn't, governor. He said he wouldn't hurt me by doing it."

"Quite right of him," Lord Beranger falters through the tears that roll down his cheeks. "Hargreaves, come closer."

He draws closer and kneels down beside the couch, and taking up one long, glittering tress, he puts his quivering lips to it.

"You may kiss me, Hargreaves," Baby murmurs, with a half smile on her pale lips. "There are no *convenances* where I'm going!"

He rises from his knees and, bending over, kisses her for the first and *the very last time.*

"Good—bye—all!" she gasps. "I have—had—a—jolly—time—but—I'm—not sorry—to—go! Go—od—bye!"

Her eyes close, a grey hue runs round the pretty lips and the shadow of the Angel of Death falls on her little face.

Only a few hours more and Baby is gone!—gone with her smiles and her wiles, her coaxing ways and her naughty ways—gone to that land which only faith can pierce and where only love can follow.

There is not a dry eye in the household, when with awesome spirit and noiseless tread they go in to see the last of her.

She lies like an exquisite waxen image, her sweet voice silenced, her blithe laugh hushed, her slender white arms crossed on her stilled heart, and a snowy Eucharis lily resting upon her breast.

"Oh, my lord! put this somewhere near her from *me*!" poor Hargreaves had said through blinding tears.

"One touch of nature makes the whole world kin," and Lord Beranger, knowing with what a true, honest, unselfish love this young fellow had loved his lost child, places the lily on her breast with his own hands.

* * * * * * * *

The day after Baby is laid to rest, Hargreaves is found near the Beranger vault; one hand grasps a locket with a bit of golden hair in it, near the other hand is the revolver with which he has shot himself. It was true what he had said, that he loved her with the love of a dog, that would just lie down and die beside her grave.

But the matter is at once hushed up, for the *convenances* do not allow of *canaille* even *killing* themselves for the sake of daughters of Belgravia.

# CHAPTER IX
# LET THE DEAD PAST BE BURIED.

"Let this be said between us here,
One love grows green when one turns grey,
This year knows nothing of last year,
To-morrow has no more to say to yesterday."

"THE pomps and vanities and sinful lusts of the flesh" being put a stop to by poor little Baby's untimely death, Lady Beranger has elected to mourn in sackcloth and ashes among the sylvan shades of Sandilands. It would be dreadful to assert that this worldly mother does not lament to a certain degree the gap in the domestic circle, or that now and again the memory of Baby's sweet pretty face and winsome, kittenish ways does not bring a mist into her fine eyes, but this much is true, that she leaves Belgravia with regret, especially as the season is not quite dead. And now that three months have nearly gone by since

"MIRABELLE BERANGER,

Aged 17,"

went away to the angels, Lady Beranger, knowing that mitigated affliction in the shape of jet and bugles are always becoming, has "just one or two intimate friends" come down to share the quiet of the country and to sympathise with the family woe.

It need not be said that, with that worldly wisdom that looks sharp after its own interests, these intimate friends are Lord Delaval and Mr. Stubbs.

Of course such glittering fish must not be lost sight of before they are safely landed.

It is not unusual in the Upper Ten, as has recently been proved, for the *noblesse* to rise from the funeral baked-meats to sit down to wedding-cake.

Anyway, as the *convenances* are not rigid on this score, it is on the cards that before Trixy's crape grows rusty she will don the orange and myrtle.

And now that Sandilands offers no flirting material with which she can keep her hand in and show off her power, save "poor Mr. Stubbs," she goes with less reluctant feet towards the altar of Moloch than she did in

Town, where her "future" cut such a comical appearance among the golden youths that she really hated the very sight of him.

"It's rather a bore that one can't go and get married respectably at St. Peter's," she remarks pettishly to Zai. "I might as well be a housemaid, to walk across the garden path to that paltry little church, and hear old Boresome gabble a few words by which Stubbs and I shall be made—*one*! Ugh! Do you know, Zai, I expect we shall be very much *two*! We haven't a single idea in common, and only one pleasure—contradicting one another."

"Don't marry him, then, for goodness sake, Trixy! You'll be a wretched girl if you do. If you *can't love* a man, you must at any rate respect him, or look up to him as having a superior intellect to your own," Zai replies, thinking of Lord Delaval; then she frowns and chases away the thought of him as fast as she can.

"Well, I don't love Stubbs—(he asked me this morning to call him *Peter*, but I *couldn't*, I really *couldn't*)—and I don't respect him particularly, and I certainly don't consider his intellect superior to mine, but I intend to marry him all the same. Love and respect! Good heavens, Zai! Such things are all very well in their way, but you don't suppose that I should think of balancing them with that lovely suite from Jackson and Graham's? Why, those white and gold chairs, with the crests carved on the backs, are ten times more worth having than all that fiddle-faddle of love and respect!"

Zai does not answer. She knows, perhaps, that some of Trixy's notions are unanswerable, and is simply conscious of the fact that she rather envies her her sentiments.

"And what's the good of having *point de Venise* on my dress for the gardeners and stable boys to gape at?" Trixy goes on, peevishly. "I think it is too bad to be done out of everything like this! I had made up my mind to have a fine wedding, all the good-looking men in town, a lot of bridesmaids, and—why, what's the matter, Zai?"

The matter is that Zai has allowed a sob to break in on her talk.

"Nothing," she says, in a low voice; "only your speaking of bridesmaids made me think of *Baby*!"

"You were always a wet blanket, Zai. Whenever one is trying to look on the bright side of things, you are sure to say something horrible," Trixy replies, in a tone of martyrdom. "*I* think of Baby too; but I drive away the thought because it is my bounden duty. Mamma says I'm not to make myself ugly with crying and fretting, and, Zai, do you know, I don't think there's much to grieve about Baby. *She's escaped marrying a—Mr. Stubbs!*"

It strikes Zai again that Trixy's ideas are a little out of the way, and wiping her tears, she takes up a book.

"I say Zai! I want to tell you something," Trixy announces suddenly, in a half whisper. "It's a secret, a dead secret, and you will have to swear you will keep it."

"I promise," Zai answers quietly, wondering what important thing is to be divulged, as Trixy crosses the room and comes close up to her.

"No, no! you must swear."

"I never swear; but my promise holds as good."

"Well, then, listen. Gabrielle told me this morning that there is something between you and Lord Delaval."

"Well, if there is, what of it?"

"Only that Gabrielle went down on her knees on the damp grass, and swore (she swears awfully, you know) that if he married you, she would destroy herself, body and soul!"

"I am sure she is welcome to him if she wants him so very much," Zai flashes impetuously; "but I *must* say that if Gabrielle really fancies he is going to be her brother-in-law, she ought to curb her feelings for him!"

Trixy opens her big blue eyes wide with amazement.

"You don't mean to tell me, Zai, that there is the very least bit of foundation for Gabrielle's fancies?"

"Yes, I do," Zai blurts out, "a very great deal of foundation. I have been engaged to Lord Delaval ever since the State Ball, and I suppose I shall marry him some day."

"And you really accepted him in cold blood, although you have always said you disliked him so?"

Zai reddens to the roots of her chesnut hair.

"Women are allowed to change their minds, I suppose?"

"You didn't change your mind, Zai. You have only accepted Lord Delaval out of pique. It's all because that dishonourable fellow, Conway, pitched you over for Crystal Meredyth. Oh! Zai! *cannot* you arrange to be married the same day as I am? It would make me so much jollier to know I had a fellow-sufferer! It is quite a month to it—lots of time to gallop through the trousseau—and then people won't say that you only married Lord Delaval when Carl had put a Mrs. Conway between you and him."

Zai looks up at her sister rather piteously; her grey eyes are dimmed with tears, her face is very pale, and there is a falter in her voice as she asks:

"When is Mr. Conway's wedding to be?"

"Just six weeks hence."

A pause. The September sun shines down hot and glary, but under its broiling rays Zai shivers. Her heart is cold, her hands are cold, and it seems to her that life altogether is awfully cold. Still in this moment she makes up her mind.

"All right, Trixy!" she cries, in ringing accents, just as if she was as blithe as the sunbeams and the birds; "the same day shall make us both—wives—on two conditions. One is that you will not tell Gabrielle a word about our little arrangements until I give you permission. The other condition is—— " She pauses a second and turns away her face, and when she speaks again her voice is so husky that Trixy wonders—"that you will never mention Mr. Conway's name to me again! Before I marry Lord Delaval, I should like to bury my dead past for ever and for ever out of sight."

"But Mamma must know of our arrangement, and she will tell Gabrielle, of course."

"Oh, no, she won't; not if I ask her. Look here, Trixy. We are a set of paupers! Even our mourning for Baby—" in spite of her she falters—"is all on credit. I heard May's man say 'Crape's a very dear article, my lady; and the deeper the affliction the more it costs, in course! So it's only the quality, my lady, as can *really* indulge in mourning; the commonality mourn usually in narrow frills or small pleats, *but the quality, to be fashionable, must mourn in deep kilts.* Sorrow cannot be better shown than by as little silk as possible, and full crape draperies, the buttons to be covered in crape, in course, and crape collars and cuffs, and jabot on the bodice.' 'The mourning must be deep, of course. I suppose, in your very large way of business, you do not trouble to make up the account but once in a year or so, do you?' Mamma asked, in her most benign voice. 'The mourning must be sent home with as little delay as possible, and of course if it inconveniences you to wait, I will give you a cheque in advance.' "

"Good gracious!" cries Trixy, "what a state of funk the mater must have been in for fear he'd take her at her word!"

"Yes; but he didn't. 'No, no, my lady. We can afford to wait quite well. We are in no hurry whatever; in fact, we shall be only too pleased and honoured by having your ladyship's name on our books, so long as your ladyship will allow us;' and it was only in this way that we got this outward and visible sign of our grief for Baby, and it is only in this way that we get our bread and butter, you know. The Governor and Mamma are delighted at your marrying Mr. Stubbs, and the idea of my catching Lord Delaval has filled their cup of bliss to the brim; so they won't do anything to make us turn rusty. Besides, Mamma knows better than to tell Gabrielle anything, in case she should put a spoke in my wheel of matrimony. She is so much in love with my *fiancé*."

"And does he care for her?"

"What a question!" cries Zai, flushing a little. "Now is it likely that he should want to marry me if he cares for my step-sister?"

"*Cela selon!*" Trixy replies carelessly, "Men don't much mind that sort of thing. I heard Charlie Wentwaite only made love to Virginia South because he admired her mother!"

"You shouldn't listen to such things, Trixy. Lord Delaval may have talked nonsense to Gabrielle, because she encouraged him, but I am sure he only cares for *me*!"

"And you—are you in love with him?" Trixy asks in a solemn voice, putting her hand on her sister's shoulder, and staring at her fixedly.

But Zai cannot or will not meet this enquiring gaze.

She springs up from her chair and throwing up the window sash looks out on the fair world, the glowing fragrant roses and the clear blue sky overhead. There isn't a fleecy cloud on the azure surface. Somehow all these things have a subtle charm of their own, and bring her an impetus to bury her dead past as fast as she can, and to begin a new era. So instead of answering Trixy, she plucks a rose with a deep blood-red heart and flings it deliberately at somebody who is lying his full length of six feet two inches on the sward, his straw hat thrown aside, and the daylight falling full on his very handsome blond face. His lids are closed, and he looks the picture of laziness—but a picture that most women would take the trouble to look at several times. As the rose falls full on the tip of his aquiline nose, he slowly opens his ultramarine eyes, and looks up at the face at the window with a depth of admiration and tenderness in the look that makes Zai blush and hastily withdraw her head.

"Yes Trixy!" she cries with quite a beaming smile. "I believe I *am* in love with him, anyway I intend to be directly I am Countess of Delaval!" And five minutes afterwards Trixy sees her on a rustic bench under a big elm tree, and Lord Delaval lying at her feet. Trixy watches them a moment. What a handsome couple they make. She sighs as she looks at them, and rather envies Zai the good looks of her lover. Then she turns away and murmurs in a tone of resignation:

"A handsome man always wants worshipping, while I like to be worshipped myself, and another thing, poor old Stubbs won't ever make me jealous!"

Made in United States
Orlando, FL
04 November 2026

90205470R00046